Hans J. Berndt

Measurement & Control
using Smartphone & Tablet

BASIC and more in a pocket

III

ESP8266, Digispark, Arduino connected to Android & Windows by TCP/IP, Wi-Fi, Bluetooth, USB, RS232

Interfaces, Values, Charts, Applications - mobile

Text © 2017 Hans J. Berndt

English translation of the original German:

„Messen Steuern Regeln mit Smartphone und Tablet"

PREFACE

This book tries to outline new ways for low-cost measurement and control solutions, using smartphone and tablet. It focuses on the interplay or interaction of portable hardware connected by means of serial connections like Bluetooth (RX/TX) and Wi-Fi (TCP/IP).

An Android smartphone or tablet is assisted by little helpers to enable finding individual wireless and mobile solutions. These are discrete hardware modules as well as software modules using other (portable) devices. As an example, a low prized Windows tablet is used as a helper, to run various available free software or applications to achieve the set goals. Free Smartphone apps are used as well.

In this book there are three sections:

The first section mainly deals with the hardware elements ESP8255 (Wi-Fi) and Digispark (USB and RX/TX) to provide low cost mobile measurement and control hardware. The usability and programmability of the ESP8266 module is dealt with three different approaches.

ESP8266BASIC programming and measurement in a browser offers completely new approaches and solutions, when it's about device-independent measurement and process control by the hobbyist or anyone else.

The second section is about the concept and application of the free software to be used on the devices to achieve the set goals as a set of tools. Both, application software and scripting languages are being used. To recognize the functionality and make adjustments to given examples or templates, if necessary, low general programming skills are advantageous.

The third section treats arbitrarily chosen concrete examples with combinations of these components. In addition to customary measurement configurations and their structure, especially the routing of signals or measured values is handled. This transfer might sometimes look a little weird; it aims to illustrate the possibilities only.

Examples include a GPS module that routes data to a smartphone via TCP/IP (Wi-Fi) or a smartphone using speech recognition and sending the result as text to a Windows tablet which in turn speaks this text back into the microphone on the phone – a kind of Chinese Whispers. Further applications show how to create live charting of measurement data in a browser using JavaScript, ESPBasic and HTML. Finally measurement values are caught without the need of the Internet using HF-reception.

To simplify and clarify the possibilities of the components or the interaction in a larger context, some schematic blocks are used.

This book is designed to complement or extend the two (German) eBooks "Measurement using a Smartphone" and "Measurement and Control using a Smartphone." A Galaxy GT-7000 Phone (Android) and a Dell Venue8 Pro Tablet 3845 (Windows 8.1) mainly were used.

Table of Content

1 HARDWARE ELEMENTS

There are several small hardware components in use to solve problems concerning the title of this book. The use of these components is mostly determined by their possible connections and functionality. First are given brief characteristics of the little hardware modules, however the main focus is pointed to the *Digispark* and the *ESP8266*. The first is called "the smallest Arduino" and lives on a tiny board with USB plug included. The ESP8266 is used as it is faster with more memory compared to an Arduino Uno. But more than this the wireless capability with the resulting opportunities is the main reason for this choice. There are three different ways to program an ESP, as described below and that is why it is universally suitable for mobile use.

DIGISPARK

The board comes designed as a USB connector and has a very small ATtiny85 microcontroller with the following components and specifications: the controller, a voltage converter with a 5 Volt output, a power LED and a controllable LED and there are 6 x I/O connections.

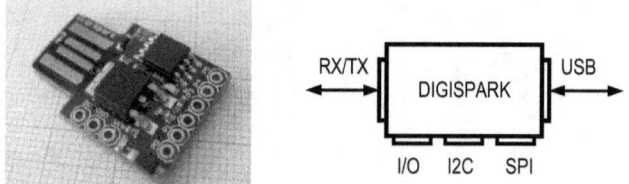

Figure 1: Digispark block and real board

Pins	Digispark
P0	D0/PWM0/AREF/MOSI/SDA
P1	D1/PWM1/MISO
P2	D2/A1/SCK/SCL
P3	D3/A3/USB+
P4	D4/PWM4/A2/USB-
P5	D5/A0

The pins can be used depending on the configuration as RX/TX (TTL-Serial), SPI, I²C, USB and ADC. Clock: 16.6 MHz, the board is tiny and has a tiny price too.

ESP8266

This component is influencing the scene of IoT since 2014. The situation at that time was: Very little documented and only accessible by AT-commands. The chip became very popular and in the year 2015 the Arduino-IDE (core) got available. In 2016, however, the simplest of all programming languages was launched as ESP8266Basic. This BASIC is a compiled sketch which once must be uploaded by a flash application. Then the programming is done in a browser and the interpreter as well as the program files of the ESP file system remains in the ESP. It is fast (80 MHz), has a lot of RAM and is native equipped with Wi-Fi. The price is very low.

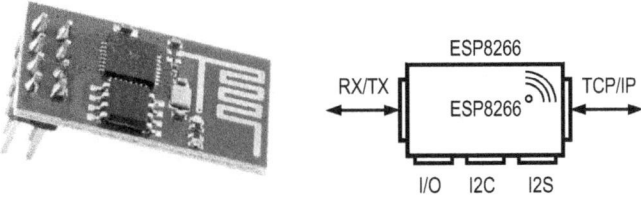

Figure 2: ESP8266-1 and block diagram

The ESP8266-01 pictured here was and is the first in the 8266 series of *expressif*. The section ESP8266-AT uses this hardware and displays applications that use the Arduino as an agent of AT-commands. Due to the rising popularity of the board and its various designs, both hardware and software also were simplified.

http://www.esp8266.com/wiki/doku.php?id=esp8266-module-family

Since the Arduino development environment (IDE) supports the ESP8266 with its core, the program development is as simple as using an Arduino, and even most sources and libraries are compatible. A sketch compiled under the core is the BASIC interpreter for the ESP8266 and this simplifies home automation again dramatically as the ESP can be programmed easily and wireless by writing a few lines. The BASIC programs are stored in the ESP. The HTML/*JavaScript* support provides a new and easy way to create individual applications. The hardware in the sections ESP8266-core and *ESP8266Basic* is an ESP BASIC WIFI Development Board (Witty-Cloud) and comes with an ESP826612F. This low-cost and easy to use board has a RGB LED and a LDR on board, the attachable USB/Serial adapter, making this module behaves like a fast Arduino including Wi-Fi.

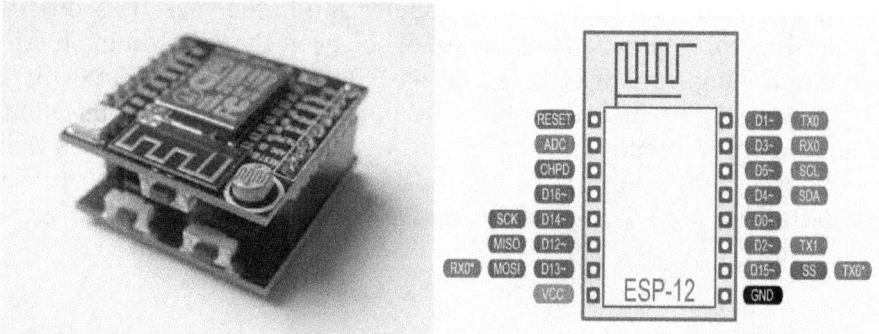

Figure 3: ESP8266-12F - "Witty-Cloud" and connections ESP-12E/F from Source: esp8266.github.io/Arduino/versions/2.0.0/doc/esp12.png

HC-06

A Bluetooth2 module (TTL RX/TX) which transmits serial data via the 2.4 GHz according to the SSL standard and with four main terminals (RX/TX/GND/VCC) a very straight forward thing. The module connects smoothly and easily to mobile Android devices, Windows, however, sometimes is a bit more reserved.

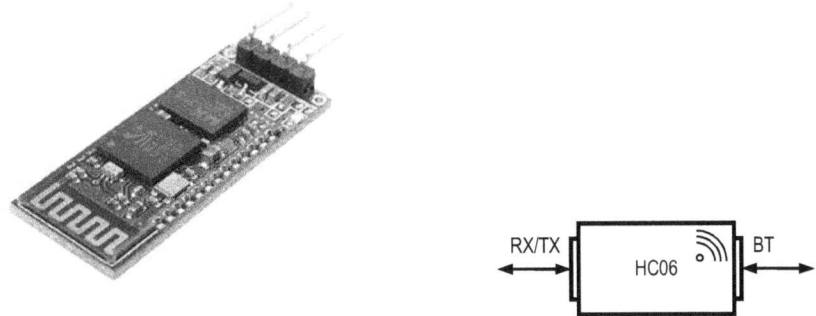

Figure 4: HC06 Bluetooth module and block

FTDI-ADAPTER (TTL)

This is a mediator between USB and TTL-RX/TX (Serial) with an USB port on the adapter which appears as a serial interface in an operating system. The voltage levels are TTL compatible and thus able to connect directly to the controller block. RS232 devices use different voltage levels and therefore cannot be connected directly. The Arduino Uno has such an adapter on board to be correspondingly easy to program.

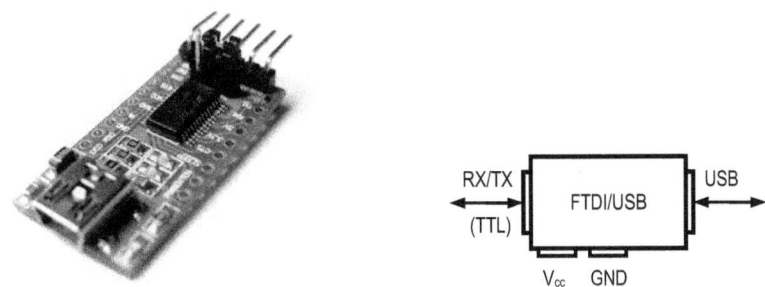

Figure 5: FTDI FT232RL USB to TTL Serial Converter and block

FTDI-ADAPTER (RS232)

This is a mediator between USB and RS232. With USB port on the adapter which appears as a serial interface in an operating system. The voltage levels are RS232 compatible and thus with these peripheral (+/- 12 V) voltage levels configured. The DB9 connector fits into the socket of serial devices that were previously usually connected to the computer (printer, mice, measuring devices). Adapters using a FTDI chipset are recognized by Android and even under current Windows versions a driver search is usually not necessary. This adapter was used in [2].

Figure 6: DIGITUS RS 232 USB Serial DB9 ADAPTER and block

ARDUINO UNO

Mainly only the programming environment of the Arduino (IDE) has its part in this book when writing software for ESP8266 and Digispark. The Arduino as a hardware element may act as agent of measurement and control data. This extremely popular microcontroller takes over control tasks in the section of ESP8266 AT when it is addressed via AT commands. Literature on this hardware is present in abundance, thus here further explanation is omitted.

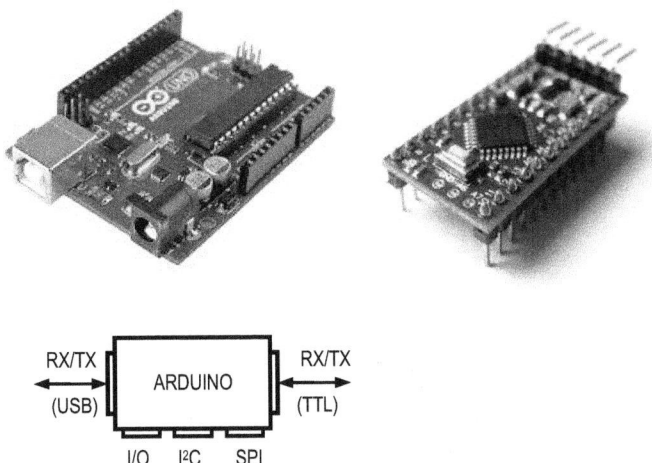

Figure 7: Arduino Uno R3 and Mini Pro and block

1.1   ESP8266 (BASIC)

Programming the ESP8266 is explained chronologically in reverse order, so ESP8266Basicfile is first.

Figure 8: Hompage ESP8266Basic

As a last resort, but the simplest kind of programming, in 2016 *ESP8266Basic* arrives, to improve the treatment further simplified tremendously and even open up new possibilities, since the transfer of sketches is omitted and everything - on top of all - works wireless. Because the BASIC is running as a server, multiple clients can access by browsers. The operating system running the browser does not matter - however - HTML5 support is desirable.

The file system and the Web server from ESP8266-CORE in the next chapter open up a lot of possibilities. "MMIsCool" had probably the same idea and posted a full Basic. Programming is done via WLAN in a browser. Any Wi-Fi device, with browser and JavaScript, can serve as programming environment. For the C programmer might be interested in the fact, that the complete source code is open source and can be compiled using the current Arduino ESP cores of the IDE. A step to personal ESP Basic?

[EDIT] [RUN] [SETTINGS] [FILE MANAGER]

ESP BASIC 3.0.Alpha 55

Device MAC: 62:01:xx:xx:97:1D

At *esp8266basic.com* a transfer program - a flasher app - can be downloaded to transfer the compiled BASIC-Sketch to an ESP8266, so no Arduino IDE or C-knowledge is required. The vocabulary is based on GW-Basic and supports all ESP properties. The programming interface and the output is a browser running on the user device.

1.1.1 ESPBASIC: SETUP

Basic is flashed once by means of the program *ESP_Basic_Flasher.exe* in order to be able to use it. This executable file is located on the website in the download section and is about 1.4 MB small. After starting the flasher the software searches the ESP on a COM port. The ESP has to be connected via USB and set to the flash mode, like using a Arduino IDE. BASIC is a fully compiled Sketch and flashed like any other sketch in the initial area of memory. In addition, the transfer program allows formatting the file system, almost a format C: from the old days.

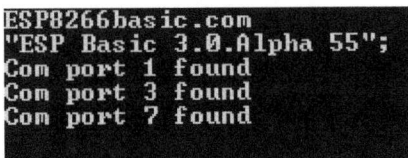

Figure 9: ESP Basic-Uploader delayed finding COM7

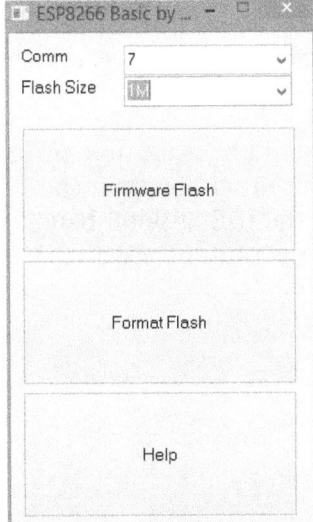

Figure 10: Upload BASIC Sketch and file system format using ESP_Basic_Flasher.exe

At *ESP8266Basic.com* this setup is explained by many pictures. Using a Witty-Cloud the setup is like this:

- Connect ESP via USB
- Start ESP_Basic_Flasher.exe (searches and finds COM7 after a while)
- Set ESP module in Flash mode
- In the flasher select COM7 and 1M
- "Firmware" (Basic-Sketch) transfer (Blue LED flashes as normal)
- Format the file system

After restarting the ESP a Wi-Fi Access Point "ESP" should be in the air and displayed in the Wi-Fi settings of the users device. Now programming can be done completely independent of other network infrastructure. Additional steps enable the ESP to connect to a router, e.g. to get access to the internet. To connect to the ESP-AP "ESP" by the browser a call to the IP "192.168.4.1" in the address field should result in the following output:

[EDIT] [RUN] [SETTINGS] [FILE MANAGER]

ESP BASIC 3.0.Alpha 55

Device MAC: 62:01:xx:xx:97:1D

Enter your own router with password in settings, save with save, restart with reset.

After a while the AP "ESP" is off air, the ESP should be accessible by its local IP from the router. Control information sent to serial output can be watched by RealTerm or the Serial Monitor of the IDE at 9600 baud at COM7. Alternatively the local IP can be found in the router software.

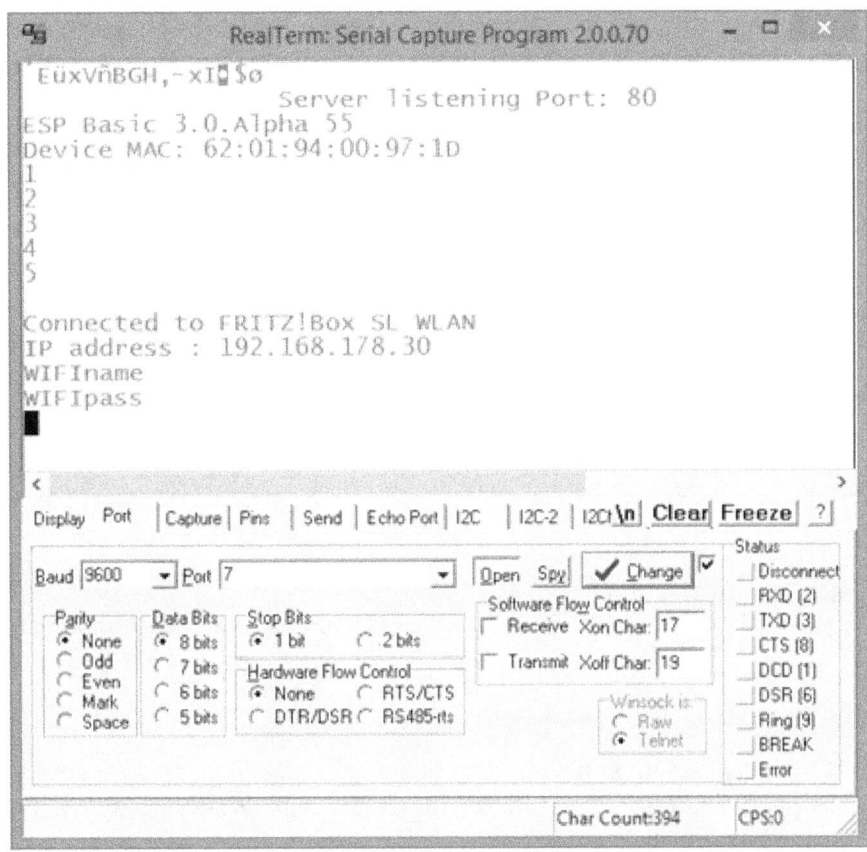

Figure 11: RealTerm for monitoring serial output of ESPBasic

1.1.1 ESPBASIC: HELLO WORLD

The first installation of version 2 of ESPBasic requires a certain amount of patience to get these words displayed. If the system is familiar, everything is straight forward and in version 3 there are many improvements. The output appears in the browser window and via the serial interface.

Three options are available:

PRINT - Output in the browser and via the serial interface

WPRINT - output only in the browser as HTML text

SERIAL PRINT / SERIALPRINTLN - Only through the serial port (9600 baud)

As of version 3 the statement *HTML* is equivalent to *WPRINT*.

In order to get the first output, you call up the "EDIT" menu from the BASIC menu and write down the program line:

```
Print "Hallo Welt"
```

In Firefox, this looks like this:

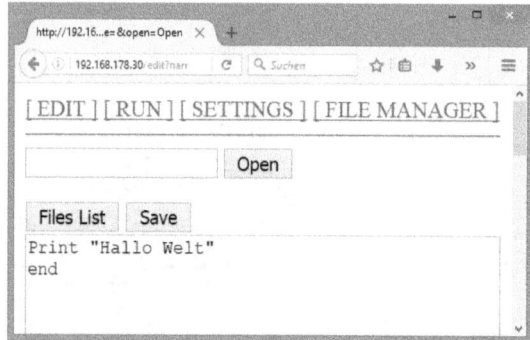

Figure 12: Programming in a browser, first ESPBasic program

Next the program must be saved by pushing the SAVE button. An empty namespace creates a "/default.bas" file. BASIC signals success with a message box "Save" to confirm.

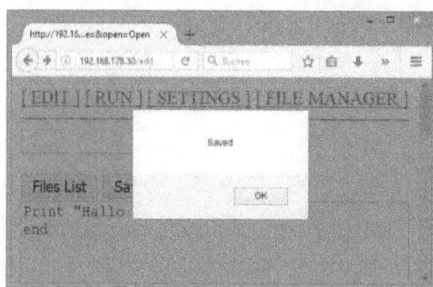

Figure 13:
First SAVE, then RUN

Now the execution is done by clicking RUN.

Figure 14: Hello World in the browser

As with *Print* both outputs are served, a terminal serial port looks like this:

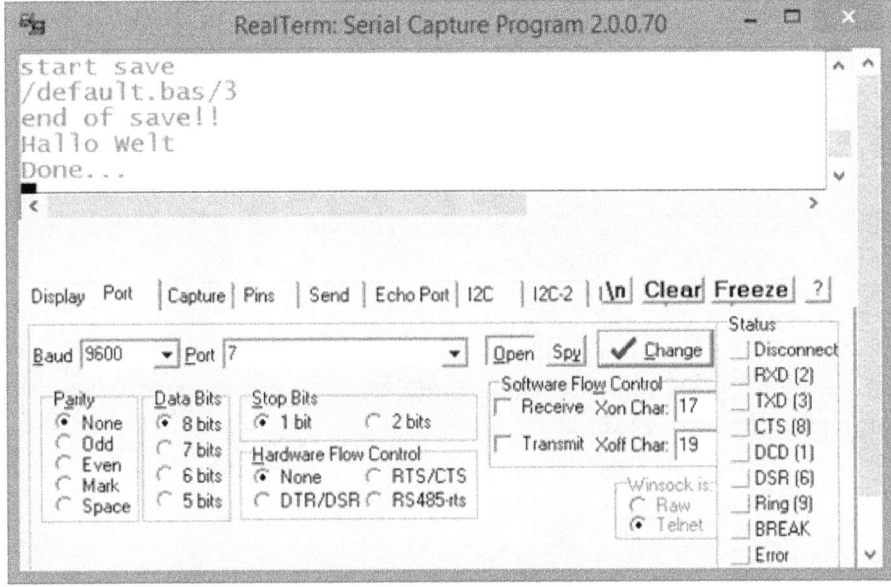

Figure 15: Hello World in RealTerm, monitoring serial outputs

These steps are required for each test that is run. In Settings the */default.bas* can be set as startup, and is started after a reset after a 30 second delay. By specifying a different name the listing is saved accordingly.

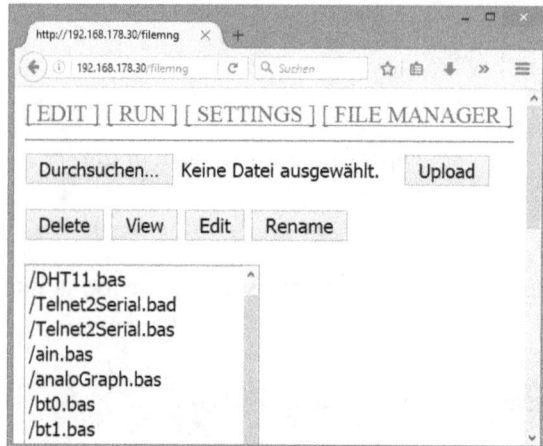

Figure 16: File Manager of ESPBasic

Using the *FILE MANAGER*, the various BASIC programs can be managed. Files can be uploaded with *UPLOAD*, for use in programs. In addition to pictures or text files is this ability interesting in terms of the use of offline *JavaScript* libraries. By the way: The entire BASIC programming interface is *JavaScript* code.

In addition to the following short samples there are more extensive examples in the third chapter.

1.1.2 ESPBASIC: BLINK

The "Hello World" for a screen output is the "blink sketch" for hardware, so the first thing is to get a LED flashing 10 times at pin 15 using the onboard-Basic of the ESP.

As described above, the BASIC Editor is available in a browser. There the following BASIC lines should be typed:

```
for i = 1 to 10
 io(po,15,1)
 delay 1000
 io(po,15,0)
```

```
delay 1000
next i
end
```

This wonderful mash-up does its job using the IO-statement PO (Por-tOut) 10 times by the iteration of a FOR loop. In between there is the Arduino-like one-second delay to pause 1000 milliseconds. Because in the past BASIC ended with an End, this BASIC program ends with an End and that is why BASIC reports a "Done".

If the ESP is called by another device using the ESP-IP, the BASIC program starts again. This demonstrates the ability that several devices on BASIC have access at the same time and the wireless ESP can be programmed together - all this can be done without a router and the internet!

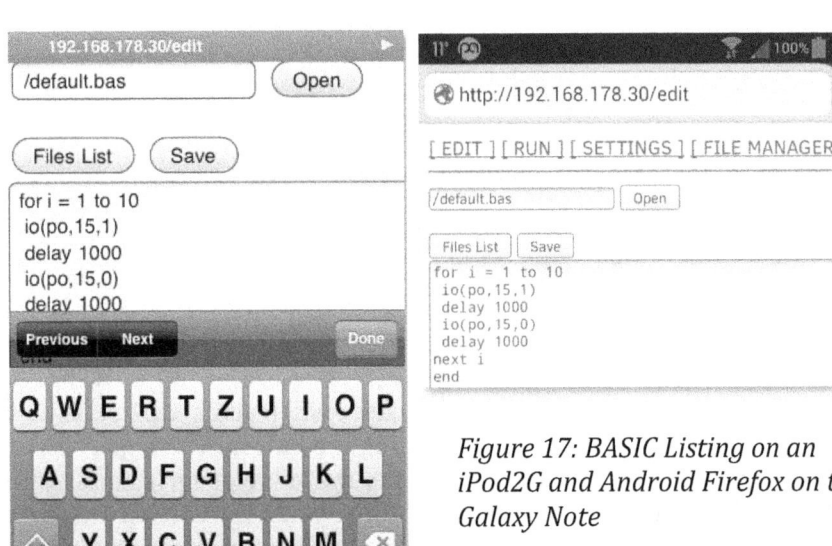

Figure 17: BASIC Listing on an iPod2G and Android Firefox on the Galaxy Note

Further instructions can be found in:

ESP8266 BASIC Help 3.0
ESP8266 BASIC Language Reference
For ESP BASIC 3.0 a XX

HTTP://ESP8266BASIC.COM
The version 3 reference contains about 60 pages with brief descriptions of all commands and instructions.

1.1.3 ESPBASIC: TIMER

ESP-Basic can be seen as a larger interaction of different components, because here classic programming with GOTO meets modern methods such as event control. A timer-controlled flashing LED in 9 lines only is the next example.

```
timer 500,[mach]
button "Exit",[ende]
wait

[mach]
io(po,15,x)
x  = not x
wait

[ende]
end
```

Using timer interval, [jump destination] the jump destination label is called every millisecond specified by the interval. A button as HTML element also includes a label which is executed when actuated. In this case, the program will be terminated by pushing the button. The *mach* section toggles the output pin 15 corresponding to the value of variable *x*. A special feature is the wait statement, which just waits for events. This type of programming structure can respond to different events, without having to constantly monitor the things in its own polling loop. Below some more examples can be found.

1.1.4 ESPBASIC: INTERRUPT

Basic provides the use of an interrupt call in a most simply way, thus polling loops are no longer required. As an example, the query of the state of GPIO pin 4 on the ESP module, which is connected to a button via a pull-up resistor, is coded. In the idle state, a HIGH is indicated at the input D2, the pushed pushbutton "pulls" the input down to LOW.

```
x = 1
meter x,0,1
interrupt d2, [wechsel]
wait

[wechsel]
x = io(laststat,d2)
wait
```

The program starts with the initialization of the variable *x*, so the HTML *meter* element first shows full scale. The *meter* for the variable *x* has a range from 0 to 1 and serves as a kind of gauge or progress bar. The Interrupt instruction ensures that status changes at D2 *ESPBasic* invokes the lines at the label *wechsel*. In this case, an interrogation of the last read state of D2 using *laststat* occurs. Accordingly, the state of the meters will change, since both sections end up with a *wait* and thus BASIC is waiting for events.

Figure 18: Monitoring a switch-state by interrupt

In Opera the result is a yellow bar as idle state. If the button is pressed, the bar will appear colorless. The SAVE-message in Opera reveals the JavaScript editor.

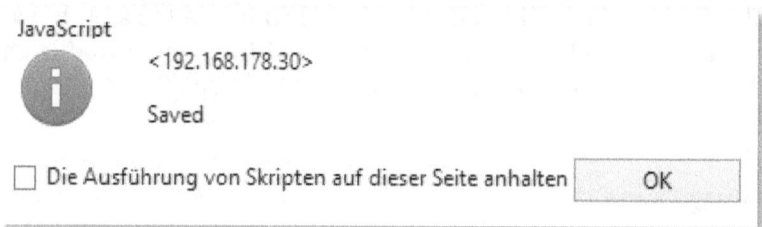

JavaScript

<192.168.178.30>

Saved

☐ Die Ausführung von Skripten auf dieser Seite anhalten   OK

Figure 19: Save and run in a tablet Opera

1.1.5 ESPBASIC: ANALOGUE INPUT

The ESP has an analogue input to which a light-sensitive resistor LDR is connected in this module *Witty-Cloud*. Thus by means of this, changing analog values can be produced in a simple way.

The inquiry will be accomplished using a timer and the BASIC Element *meter*. The 10-bit resolution of the ADCAI results in a range 0 to 1023. BASIC results even provide values up to 1024. The analogue value of the *IO* instruction *AI* (AnalogIn) is assigned to a variable *a*. Since the meter element is linked to this value, the *meter* follows the analog value and the timer query routine is performed every 100 ms. Current measurement is signaled by the blue LED blinking to get visible hardware feedback of the running Basic.

Here are the lines:

```
x = 0
timer 100,[mach]
meter a,0,1000
button "Exit",[ende]
wait

[mach]
io(po,15,x)
x   = not x
a = io(ai)
wait

[ende]
end
```

As a small hoax, the *meter* is only assigned to a maximum value of 1000, which results in turning the green bar in Opera to yellow if the limit value is exceeded.

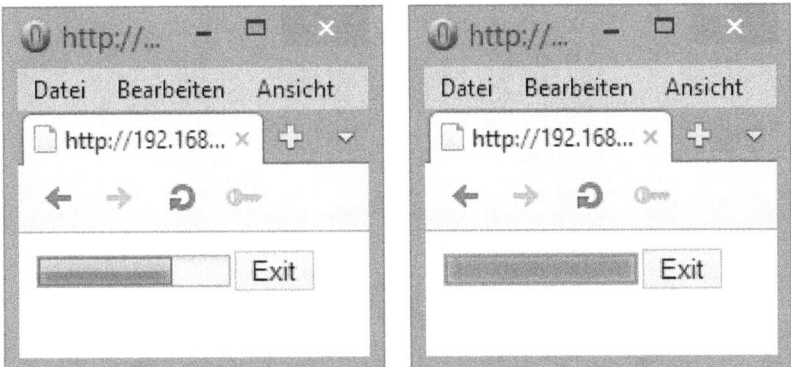

Figure 20: Analogue display of the ESP8266F-ADC (right is yellow)

Assigning a variable to a *textbox* is the best way to show its changing value as straight text. The extended listing with numeric output consists of one more line:

```
x = 0
timer 100,[mach]
meter a,0,1000
textbox a
button "Exit",[ende]
wait

[mach]
io(po,15,x)
x  = not x
a = io(ai)
wait

[ende]
end
```

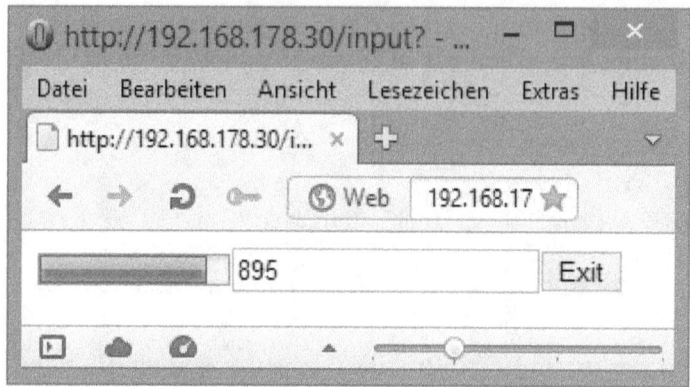

Figure 21: Analogue display and numeric value in a textbox

1.1.6   ESPBASIC: LISTING VALUES

A simple listing of analogue measurement values with time stamp can be output directly in the browsers window using Basic. In order to obtain a decimal point number as comma separated (German) number, the point is replaced by the *replace()* function. The same values can be copied on/to different devices if there are smartphone or tablet at the same time.

```
t/s       PIN ADC
0,012      157
1,013      975
2,011      675
3,011      226
4,012      397
5,016      857
6,016     1024
7,015     1024
8,015      714
9,015      370
10,015     200
Done...
```

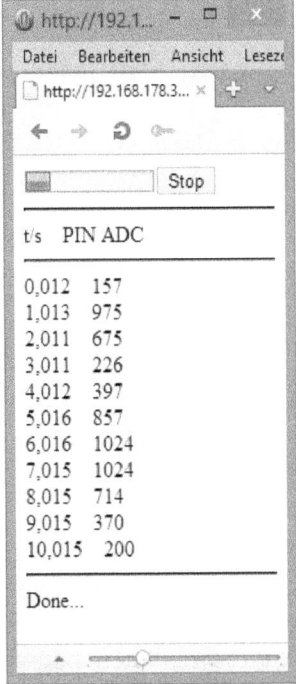
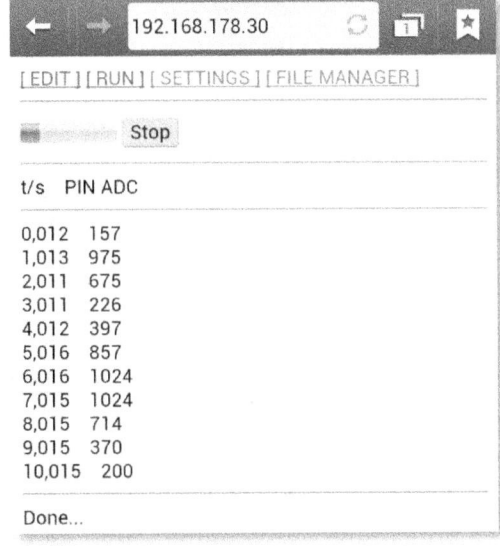

Figure 22: Measurement values in Opera and in Android stock-browser on Galaxy Note

```
x = 0
meter v,0,1000
button "Stop",[ende]
wprint "<hr>t/s    PIN ADC<hr>"
timer 1000,[machmal]
t0 = millis()
gosub [mach]
wait

[mach]
io(po,15,x)
x = not x
t = ((millis()-t0)/1000)
v = io(ai)
s = replace(str(t),".",",")
wprint s&"    "&v&"<br>"
return

[machmal]
gosub [mach]
wait
```

1.1.7 ESPBASIC: DIGITAL OUTPUT

Control often is nothing but switching. The simplest case is to switch a digital output on or off.

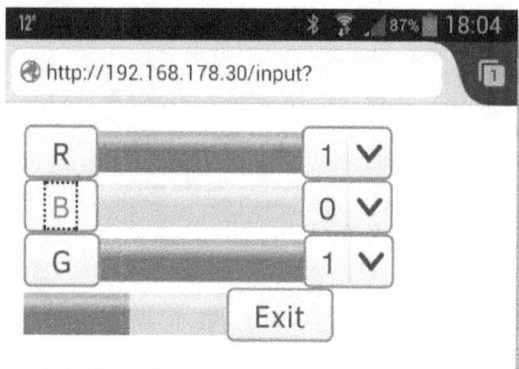

Figure 23: Switching RGB-LED of the ESP8266-module including LDR-analoguedisplay

The *Witty Cloud* board has a mounted RGB LED, which is connected to the three pins 15, 12, 13 (in that order). Using three *button* elements, the program branch is taken, or in other words, three *button* elements solve each an event, which is used to switch the corresponding output. The optical feedback in the browser takes three *meter* elements, as described earlier in the button example with interrupt. In reality, the current status is displayed by the query via *laststat*. Since transmissions via sockets or TCP/IP do not always react reliable, it is monitored by a kind of polling using a timer. For aesthetic reasons, a *dropdown* element is placed to the right of the *meter* element. This control is abused as a 0/1 indicator and may confuse the user because it does not react as an input element - the switching is done by pushing the three buttons.

```
'RGB SWITCH
x = 0
blau=-1
grun=-1
rot=-1
timer 100,[mach]
button "R",[rot]
meter r,0,1
dropdown r, "0,1"
wprint "<br>"
button "B",[blau]
meter b,0,1
```

```
dropdown b, "0,1"
wprint "<br>"
button "G",[grun]
meter g,0,1
dropdown g, "0,1"
wprint "<br>"
meter v,0,1000
button "Exit",[ende]
wait

[mach]
io(po,2,x)
x  = not x
v = io(ai)
r = abs(io(laststat,15))
b = abs(io(laststat,13))
g = abs(io(laststat,12))
wait

[rot]
io(po,15,rot)
rot = not rot
wait

[blau]
io(po,13,blau)
blau = not blau
wait

[grun]
io(po,12,grun)
grun = not grun
wait

[ende]
io(po,2,-1)
cls
end
```

For layout reasons there is an HTML *
* between the elements. Finally the blue, brighter LED at pin 2 will be turned on and the screen is cleared. The flashing of the tiny blue LED is visible as the brightness in the analogue bar pulses (LED query) next to the exit button significantly when a reflector is used. Again this simple user interface can be viewed and clicked/touched on different devices.

1.1.8 ESPBASIC: ANALOGUE CONTROL

The outputs of the ESP are PWM-capable, so they can output a kind of analogue voltage by means of pulse width modulation. The arithmetic mean voltage of a rectangular signal corresponds to its duty cycle and in this way the brightness of the three LED can be controlled depending on the output value in a range 0 (off) to 255 (on).

Figure 24: HTML-Slider control brightness

ESPBasic offers a *slider* as a control element. Programming is as easy as the *meter*, except that the associated variable is changed in the range specified and then by the slide value set by the user in that range. The *slider* moves accordingly if the variable is changed by code and the three LED at the outputs 12, 13 and 15 can now be controlled by 256 levels of brightness with the following source code.

```
'rgbSLIDE
wprint "R "
textbox r
slider r,0,255
wprint "<br>G "
textbox g
slider g,0,255
wprint "<br>B "
textbox b
slider b,0,255
wprint "<br>"
x = 0
T = 500
wprint "A "
textbox v
wprint "     "
meter v,0,1023
```

```
timer T,[mach]
wprint "<br><br>Intervall/ms: "
dropdown bla, "10,100,500,1000,10000"
button "Exit",[ende]
wait

[mach]
if or <> r then io(pwo,15,r)
if og <> g then io(pwo,12,g)
if ob <> b then io(pwo,13,b)
io(po,2,x)
x   = not x
v = io(ai)
r = abs(io(laststat,15))
b = abs(io(laststat,13))
g = abs(io(laststat,12))
or = r
og = g
ob = b
T = val(bla)
timer T,[mach]
wait

[ende]
io(po,2,-1)
cls
end
```

Again a timer polls the output state by query. The polling interval can be changed by the dropdown element and the assigned variable blah.

1.1.9   ESPBASIC: TEMPERATURE AND HUMIDITY

ESPBasic contains some (Arduino) libraries and that is why it is easy to use common hardware elements. An Example is the sensors of the DHT series for temperature and humidity with simple routines. This sensor and an Arduino can be found here:

http://www.hjberndt.de/soft/arddht11.html

Four *ESPBasic* commands support this sensor, used here in the cheap version DHT11 with only positive and integer data. The structure of the code is not very different from the previous listings, only the measured value inquiry appears according to the sensor hardware differently.

Figure 25:
From "Daten im Gänsemarsch"

```
'DHT 11 an Pin 2
DHT.SETUP(11, 2)
textbox t
textbox h
textbox i
wprint "<hr noshade size=1>"
bla = 2000
wprint "Intervall/ms:    "
dropdown bla,"2000,5000,10000,20000,30000,60000"
wprint "<hr noshade size=1>"
button "STOP",[ende]
timer bla,[messen]
wait

[messen]
t = DHT.TEMP()
h = DHT.HUM()
i = DHT.HEATINDEX()
timer bla,[messen]
wait

[ende]
wprint "Messung beendet."
end
```

Figure 26: Output on iPod Touch 2G and on a smartphone

1.1.10 ESPBASIC: ESP-APP MENU

As programs or apps can be loaded using *LOAD*, a small menu may be designed. This is somewhat reminiscent of the beginnings of the "Home-Computing," according to the prompt "Press Any Key". However, today the selection is made in a C-written BASIC using *JavaScript* buttons in a HTML5-Browser via Wi-Fi. Using appropriate filenames, a menu listing might look like this:

A	Abfrage des Tasterzustandes an Pin 4
B	Schalten der RGB-LED und LDR-Abfrage
C	Temperatur und Luftfeuchte
D	RGB mit Analog-Schiebern

Figure 27: Program selection by buttons

```
cls
button "A",[A]
```

```
wprint " Abfrage des Tasterzustandes an Pin 4<br>"
button "B",[B]
wprint " Schalten der RGB-LED und LDR-Abfrage<br>"
button "C",[C]
wprint " Temperatur und Luftfeuchte<br>"
button "D",[D]
wprint " RGB mit Analog-Schiebern<br>"
wait
[A]
load "/din.bas"
[B]
load "/rgb.bas"
[C]
load "/DHT11.bas"
[D]
load "/rgbSlide.bas"
```

The programs only need to load the file "/menu.bas" using an Exit button if the lines above are stored as that listed "*.bas" filename. A *cls* as first instruction lets things look a little clearer.

```
[ende]
wprint "Menue wird aufgerufen."
delay 3000
load "/menu.bas"
```

1.1.11 ESPBasic: Measurement Values by Mail

Measurement data does not always have to be available live and in real-time. In some cases it is better to buffer the data in order to read or analyze later. The constant connection to the *ESP8266* is not always desirable or possible. At remote locations, the measuring device may fail and locally stored readings would be lost. A solution can be storage in the network. The many possible cloud locations aside, here it will be stored on a mail server. Based on the measurement data from the previous example, the temperature, the humidity, the wind chill and a brightness value is to be sent by mail once an hour.

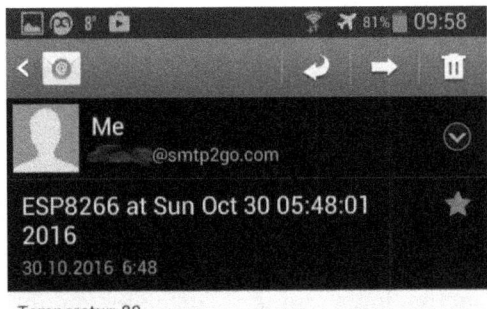

Figure 28: Esp8266 logs by e-mail on the smartphone

Only two lines of code enable sending an email in *ESPBasic*. Of course an Internet connection via Wi-Fi is required.

```
setupemail "mail.smtp2go.com", 2525, "username", "mailpassword"

email "empfänger@mail.com","username@smtp2go.com", "Betreff","Inhalt"
```

As described in the *ESPBasic* reference, a free trial account at *smpt2go.com* is needed, since those simple calls do not work for most common server accounts for security reasons. Registration requires a new user name and an existing email address and a new password to log to the site of *smpt2go.com*. After signing up you get the new email address provided by a confirmation with a request for verification. In this case, you end up as a registered user on a site where you get an e-mail password, which can be changed any time by the user according to own preferences. Finally, the dashboard opens with the activities.

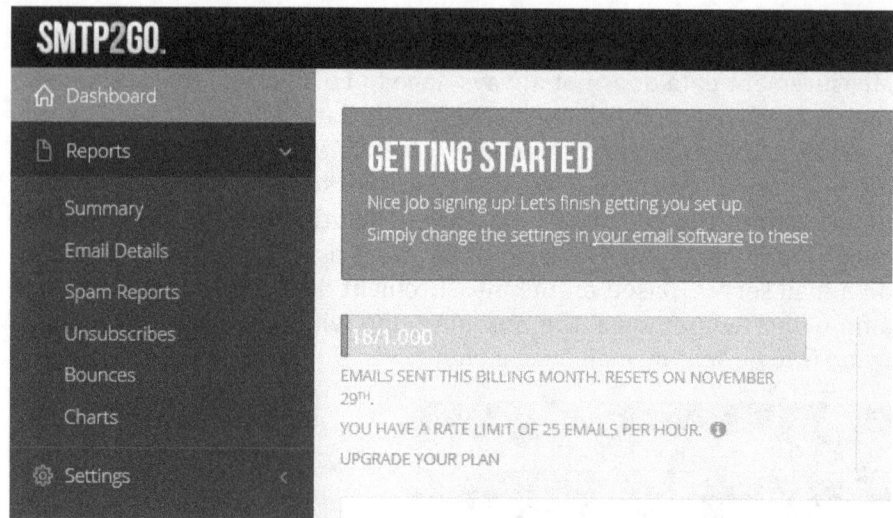

Figure 29: SMTP2Go's Dashboard of a trial account

In order to complete the process, the email address is also the receiver of the forwarded messages. Suppose the login name is *otto2go* and *otto@otto.de* an existing account, where the emails of the ESP are to be sent to, the user name at smpt2go is *otto@otto.de* too and password for login and email is *otto2000*, then a test email from ESP with the words "Hello world" is shipped by mail using following lines:

```
setupemail "mail.smtp2go.com", 2525, "otto@otto.de", "ot-
to2000"
```

```
email "otto@otto.de","otto2go@smtp2go.com", "Test-
mail","Hallo Welt"
```

The trial account only allows 25 emails per hour and 1,000 emails per month. For testing purposes and other applications this might be a nice working method, bringing all measurement and protocols by email. During a night this scenario was tested on an *ESP8266* and the below BASIC Listing sent measurement data by email until early in the morning.

Figure 30: Overnight-measurements and it's getting brighter

```
dht.setup(11, 2)
crlf = chr(13) & chr(10)
timer 3600000,[messen]
'wait

[messen]
t = DHT.TEMP()
h = DHT.HUM()
i = DHT.HEATINDEX()
l = io(ai)
msg = "Temperatur: " & t & crlf & "Gefuehlt: " & i & crlf
msg = msg & "Luftfeuchte: " & h & crlf & "Helligkeit: " &
l
print msg
setupemail "mail.smtp2go.com", 2525, "otto@otto.de", "ot-
to2000"
email "otto@otto.de","otto2go@smtp2go.com", ""ESP8266 at
" & time(), msg
wait
```

After the initialization of the sensor at pin 2, a timer is set to 60 minutes and immediately invoked when the *wait* is commented out. Next the measurement and the creation of the message in *msg* follows. With the UTC time in the subject, the electronic mail is launched. With this method, the start-up of "/default.bas" and the sleep command for the ESP long-term measurements with low energy consumption should be possible. Here, the ESP8266 is sleeping e.g. for three hours at a current consumption of a few micro amps, wakes up, connects to the Wi-Fi Station, the BASIC program starts sending the measured values and then falls to sleep. Due to *ESPBasic,* the matter remains pleasingly simple and clear.

In the third chapter more applications are presented for ESPBasic matching the context of this book.

1.2 ESP8266-CORE (WITHOUT ARDUINO)

The so-called Arduino-Core for the *ESP8266* was created due to its popularity and the possibility to integrate other hardware into the newer programming environment (IDE) of the widely used Arduino. In fact programming for this hardware is like programming an Arduino. Due to the intelligent integration of source code and libraries, Arduino projects can be transferred in part without change. Thus, the programming threshold lowered - compared to the AT variant, which requires an Arduino as an intermediary - many times. The speed and the memory of the ESP can now be fully exploited using a compiled high level language. Finally, a file system is available to handle files in the storage area as usual. The uploading of data, images, etc. is possible by means of integrated tools of the IDE. A search for "ESP8266 Arduino Core" finds the English reference for the current version of the ESP8266 Arduino core documentation or directly via the link:

https://github.com/esp8266/Arduino

1.2.1 ESP8266-CORE: SETUP

The procedure for integrating the ESP8266 into the programming environment (IDE) of the Arduino is described in the internet. An English version can be found at:

http://esp8266.github.io/Arduino/versions/2.1.0-rc2/doc/installing.html

The steps in brief:

Download Arduino 1.6.5 (not 1.6.6) from the website arduino.cc, install the IDE and run.

Under factory default Additional Board administrator URLs enter http://arduino.esp8266.com/stable/package_esp8266com_index.json file/

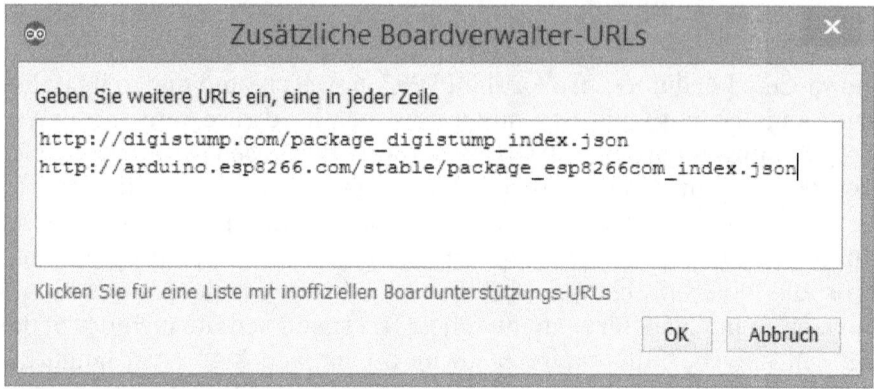

Figure 31: Board Manager of the IDE with two additional boards

Tools/Board: Select Board Manager and select the desired version and install. For this download operation an Internet connection is required.

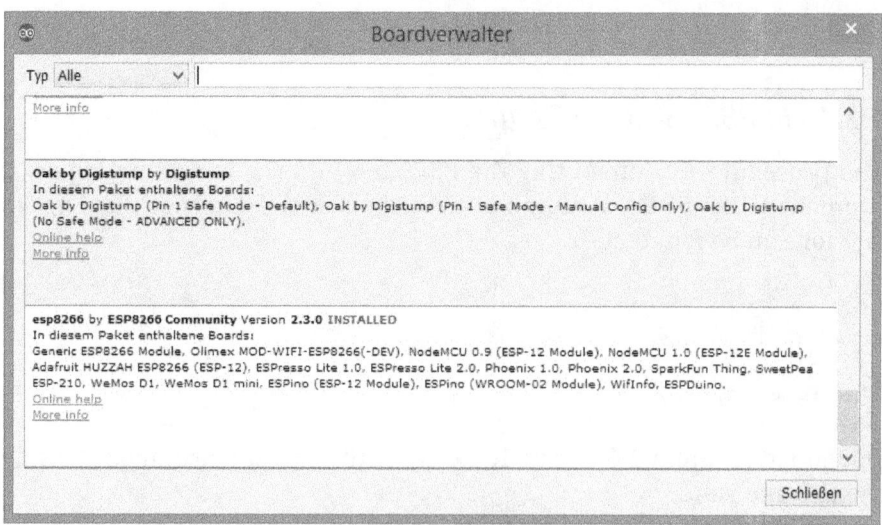

Figure 32: Installing the core for the ESP8266

Finally, at Tools Board: Select *ESP8266 Generic modules*.

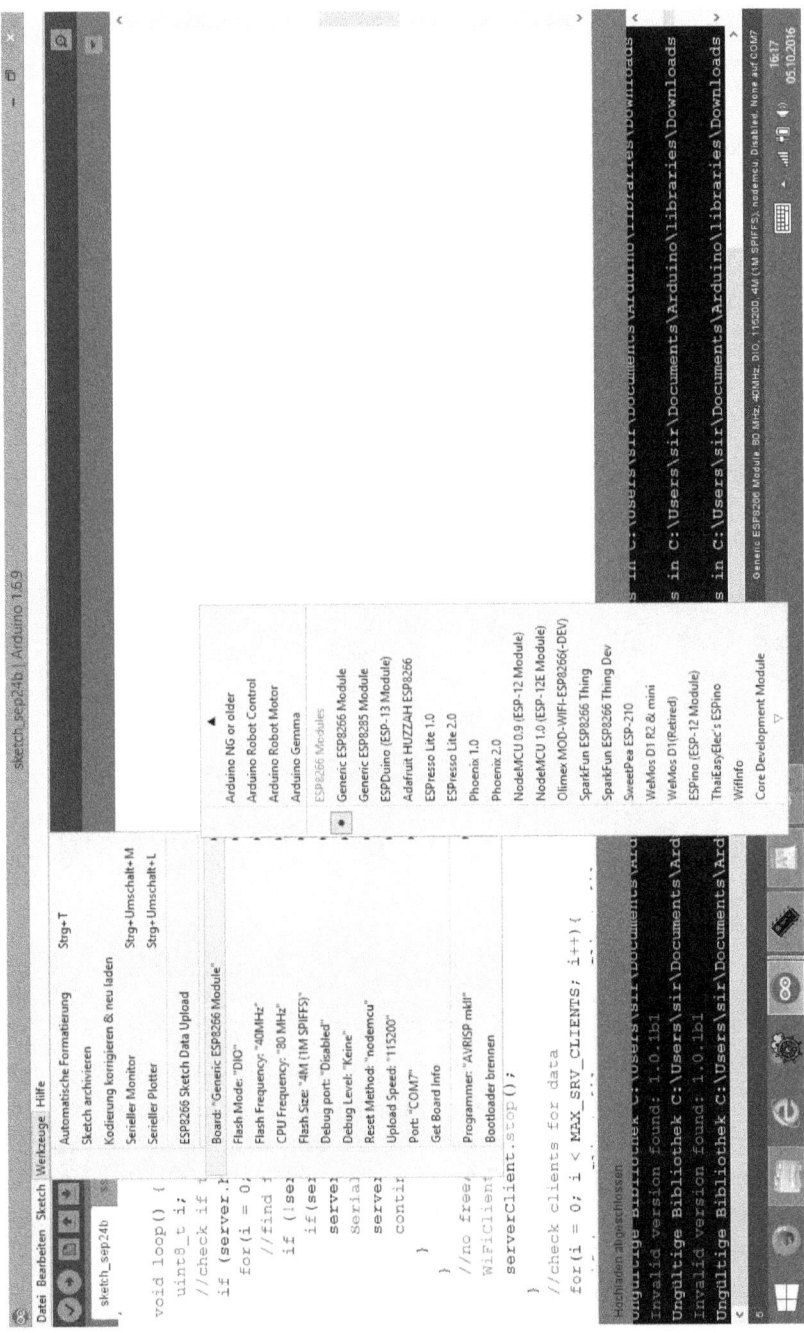

Figure 33: Board selection and examples available

Uploading a sketch using the *Witty Cloud* board with LDR, RGB LED and three buttons is done just as easily as with the Arduino, if the following settings apply:

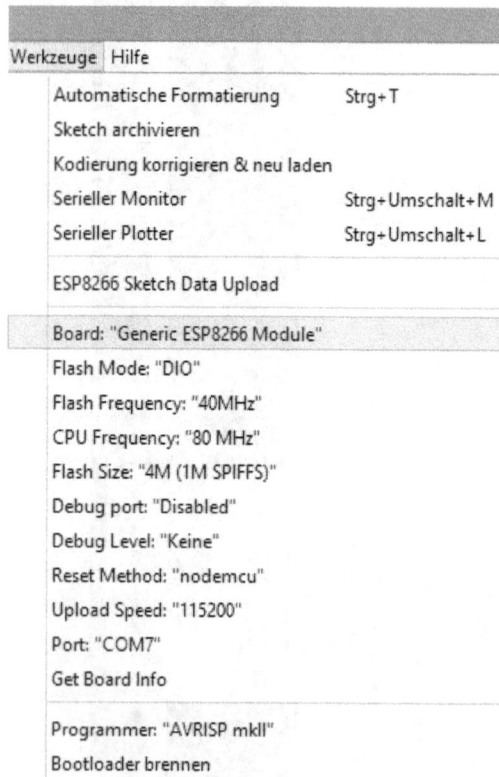

Settings for the ESP8266 module Witty-Cloud

1.2.2 ESP8266-CORE: BLINK

The standard flashing sketch is suitable for showing how the *ESP8266* has integrated into the environment. If the ESP has connected at the corresponding port (Tools) next is opening the File */examples/ 01.Basics/ blink* the original Example for an Arduino, which flashes the Pin 13 LED every second.

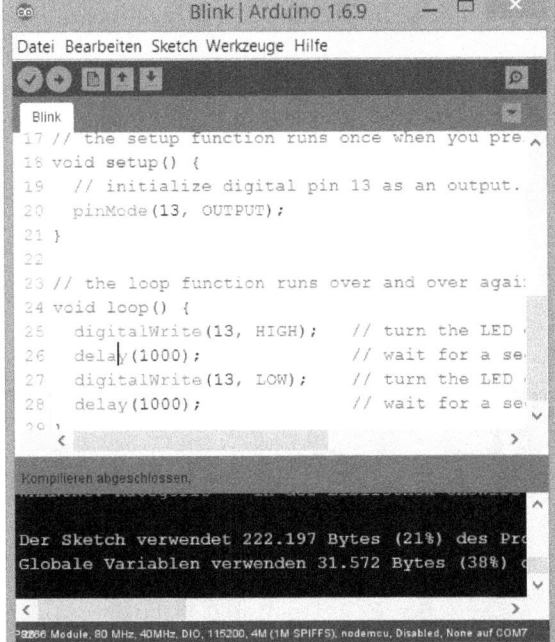

Figure 34: Blink for Arduino without any changes

After compiling, 21% memory usage is reported by 1044464 bytes. The subsequent slightly longer transmission lets the small blue LED of the ESP flash while transferring the sketch, as seen with transmissions to the Arduino. After that, the blue RGB LED connected to pin 13 should light up every second. For flashing the red LED on that board, pin 15 has to be chosen. The sketch runs without changing the Arduino source code on an Esp8266.

1.2.3 ESP8266-CORE: WIFI-SCANNER

The special ESP8266 features can be found in the examples from the ESP8266WiFi section. The unchanged sketch *WiFScan.ino* outputs at a serial speed of 115200 bd the following information to the Serial Monitor (*Tools/Serial Monitor*) in an endless loop.

There are two encrypted (*) AP (SSID) with the signal strength shown. This example integrates the library ESP8266WiFi that defines all the essential core functions. In the IDE, the example can be found as part of ESP8266WiFi templates.

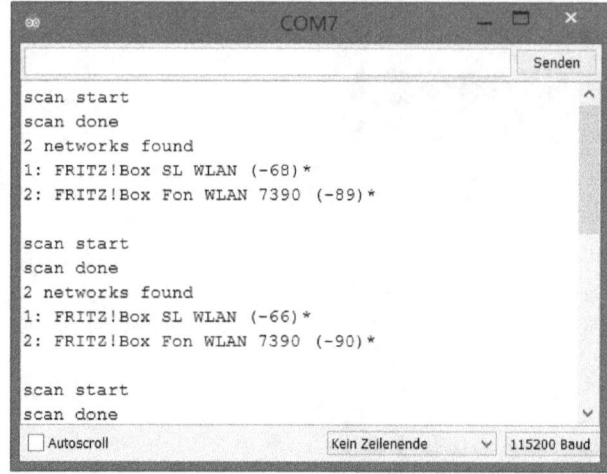

Figure 35: WiFiScan output

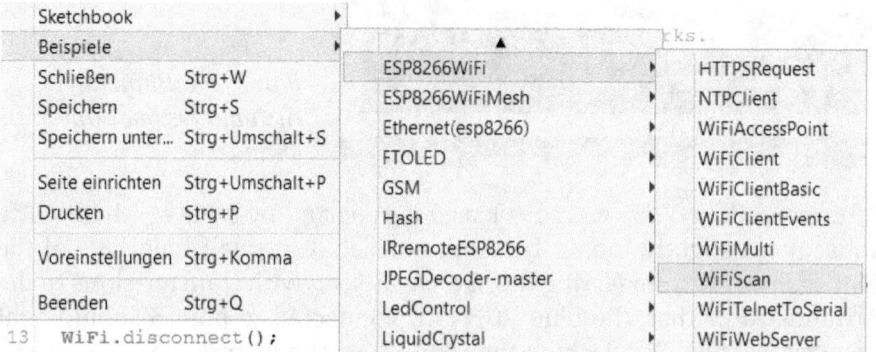

Figure 36: WiFiScan in the examples of esp8266 core

1.2.4 ESP8266-CORE: INFO

Having done these first steps, it might be useful to obtain more detailed information on the module used. At least the actual available memory size should be of interest. In the reference to IDECore there are corresponding special instructions listed with a code example. The result for this module is as follows:

Getting ESP8266 info 5 seconds from now.

Heap:	*46576*
Boot Vers:	*31*

CPU/MHz: *80*
ChipID: *45524*
FlashChiID: *1458400*
FlashChipSize: *4194304*
FlashChipSpeed   *40000000*
CycleCount   *2154190587*
Vcc/mV: *1322*

The output speed of the serial monitor is set to 9600 baud in the following listing:

```
//http://esp8266.github.io/Arduino/versions/2.0.0/doc/lib
raries.html#esp-specific-apis

#include <ESP8266WiFi.h>
extern "C"
{
#include "user_interface.h"
}
ADC_MODE(ADC_VCC);

void setup()
{Serial.begin(9600);delay(5000);
 Serial.println("\nGetting ESP8266 info 5 seconds from
now.");
 delay(5000);
 // print out all system information
 Serial.println();
 Serial.print("Heap: \t\t");
 Serial.println(system_get_free_heap_size());
 Serial.print("Boot Vers: \t");
 Serial.println(system_get_boot_version());
 Serial.print("CPU/MHz: \t");
 Serial.println(system_get_cpu_freq());
 Serial.print("ChipID: \t");
 Serial.println(ESP.getChipId()  );
 Serial.print("FlashChiID:\t");
 Serial.println(ESP.getFlashChipId()  );
 Serial.print("FlashChipSize:\t");
 Serial.println(ESP.getFlashChipSize() );
 Serial.print("FlashChipSpeed\t");
 Serial.println(ESP.getFlashChipSpeed());
 Serial.print("CycleCount\t");
 Serial.println(ESP.getCycleCount()  );
```

```
 Serial.print("Vcc/mV:\t");Serial.println(ESP.getVcc() );
}
```

```
void loop() {}
```

Flash size is 4 MB, which was set above in the board selection. There is 1 MB available for the file system. So this ESP is about a Pentium 90 with a floppy drive.

Adding the line *WiFi.printDiag (Serial)* at the end, current settings of the ESP's serial stream are revealed.

1.2.5 ESP8266-CORE: ANALOGUE/DIGITAL-PLOTTER

The latest versions oft the IDE come with an integrated serial plotter, which simplifies the fast graphical display of values. Numbers that appear in the serial monitor as text, result in the plotter output as a simple but quick graph. The integrated LDR at the analogue input of ESP8266F-Module returns digital values from 0 to 1023 by means of the on board 10 bit analogue converter. Here a short example to illustrate this.

The analogue value is output serial in the main loop as a number. The template is the standard Arduino sketch from the examples: *File /Examples /Basics /03 Analog /AnalogInOutSerial,* which looks shortened like this:

```
const int analogInPin = A0;
void setup()
{ Serial.begin(9600);
}

void loop()
{ Serial.println(analogRead(analogInPin));
  delay(2);
}
```

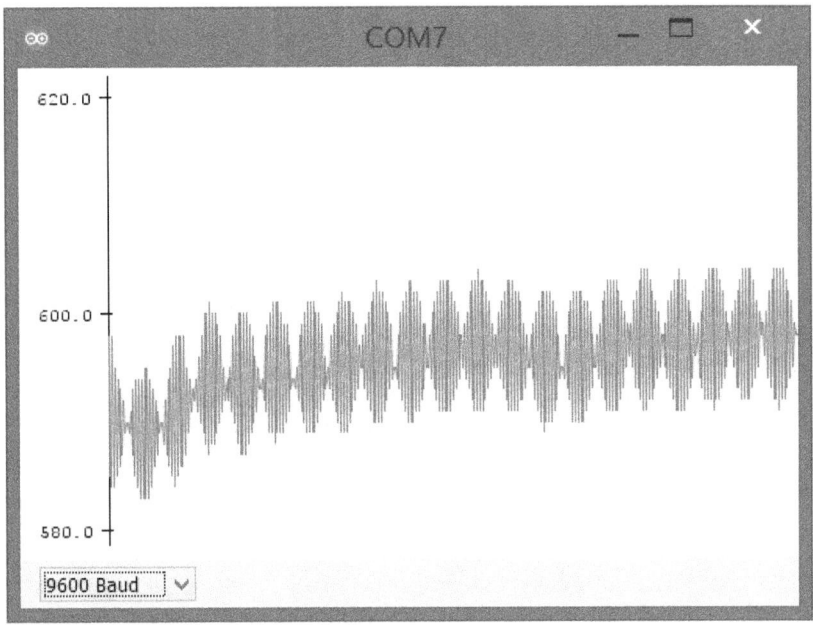

Figure 37: Brightness variation of a 50 Hz bulb

The Serial Plotter of the Arduino IDE shows brightness values coming from a daylight shielded light bulb at a detectable alternating current of 50 Hz. Taking into account the inertia of an LDR and the ADC, now is the time to experiment with the transmission rate and the delay slightly. The built-in button on pin 4 then results in the shown rectangular plot after transferring the sketch below.

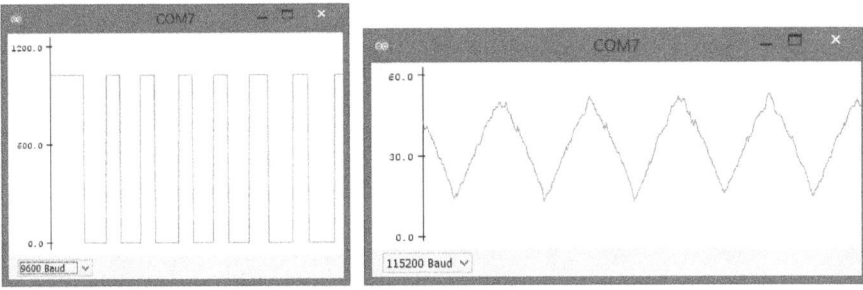

Figure 38: Switch at pin 4 and brightness plot

```
void setup()
{ Serial.begin(9600);
  pinMode(4, INPUT);
}

void loop()
{ Serial.println(analogRead(4));
  delay(2);
}
```

By iteratively controlling the brightness of an LED and a corresponding measurement of the brightness should result in a triangular waveform that can be generated in the plotter window, if the ambient light is shielded. As a reflector, a sheet of paper will do fine.

This is the resulting sketch with minor changes in the original Fade template:

```
int led = 12;

  // the pin that the LED is attached to
int brightness = 0;
  // how bright the LED is
int fadeAmount = 5;
  // how many points to fade the LED by

void setup()
{Serial.begin(115200);
 pinMode(led, OUTPUT);
}

void loop()
{brightness = brightness + fadeAmount;
 if (brightness == 0 || brightness == 255) fadeAmount = -
fadeAmount ;
 analogWrite(led, brightness);
 // wait for 30 milliseconds to see the dimming effect
 delay(30);
 Serial.println(analogRead(A0));
}
```

1.2.6 ESP8266-CORE: OPTICAL OSCILLATOR

Using the three LEDs in red/green/blue and the LDR as a light sensor, it might be possible to code an optical oscillator. The brightness of the LED is set to increase when it gets darker, thus a negative feedback. The analogue input ADC provides brightness values from 0 to 1024 (!) and the analogue output with PWM has a range of 0 to 255. A division by 4 or right shifting 2 bits, match the two ranges. The feedback is created by a subtraction. In the listing the analogue value input is y, the analogue output is x. The value of x also is sent to the plotter, to observe the result of the oscillator. The low brightness of the LED in ambient light requires an optical shield and corresponding optical coupling between LDR and the LED. A curved piece of paper again will do fine.

Figure 39: Simple reflector for feedback

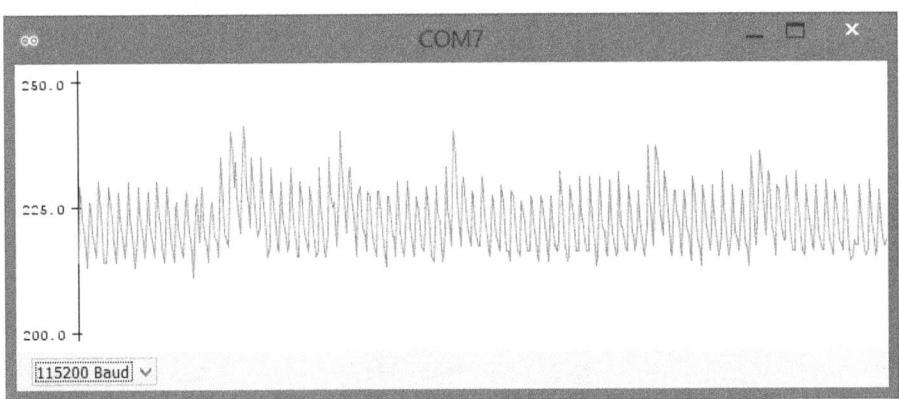

Figure 40: Oscillating brightness

```
#define GN 12
#define RT 15
#define BL 13
```

```
void setup()
{ Serial.begin(115200);
}

void loop()
{ int x=analogRead(A0);
  delay(2);
  int y=256-(x/4);
  analogWrite(GN, y);
  analogWrite(BL, y);
  analogWrite(RT, y);
  Serial.println(x);
}
```

1.2.7 ESP8266-CORE: LUXMETER

On his website, Mr. Kriwanek describes how to convert an LDR-measured value into the unit Lux by comparison. The used photo resistor is probably not identical to that of the ESP-module, yet here a light measurement in Lux has to take place. Original:

http://www.kriwanek.de/arduino/sensoren/301-arduino-luxmeter-mitfotowiderstand-ldr.html

His elaborated in meticulous detail results are embedded as formulas with minor adjustments in the following function *LDR2Lux*. The short listing of the analogue plotter earlier, is now extended with this function.

```
unsigned long LDR2Lux(int reading)
{float e = 2.718281828459;
 float x = float(reading);
 if (reading >=  0 && reading <= 155)return 0;
 if (reading > 155 && reading <= 350)  return
0.0042273988 * x * x - 1.0130028488 * x + 55.4403759239;
 if (reading > 350 && reading <= 650)return 11.7717399221
* pow(e, (0.0083003710 * x));
 if (reading > 650 && reading <= 936)return  0.3373539789
* pow(e, (0.0134529914 * x)) + 448.5;
 return 100000;
}

const int analogInPin = A0;
```

```
void setup()
{ Serial.begin(9600);
}

void loop()
{ Serial.println(LDR2Lux(analogRead(analogInPin)));
  delay(2);
}
```

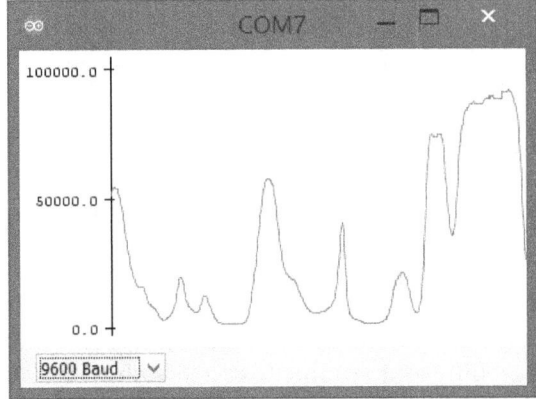

Figure 41: Serial plotter showing Lux values at the Y-axis

1.2.8 ESP8266-CORE: TFT-DISPLAY

An Arduino TFT display can be connected to an ESP. The libraries of Adafruit work fine, albeit with minor adjustments. To control a display using the chipset ST7735, the two libraries of Adafruit, and the SPI library are required. The display used here initializes as indicated in the following lines.

```
#include <Adafruit_GFX.h>     // Core graphics library
#include <Adafruit_ST7735.h>  // Hardware-specific library
#include <SPI.h>
#define TFT_CS    15
#define TFT_RST   12
#define TFT_DC    16
#define TFT_SCLK  14
#define TFT_MOSI  13
Adafruit_ST7735 tft = Adafruit_ST7735(TFT_CS,    TFT_DC,
TFT_RST);
```

The available *https://github.com/nzmichaelh/Adafruit-ST7735-Library* library works for this module and the ESP, however, due to the processors speed of 80 MHz, the graphics demo is just a short fun comparing to an Arduino Uno, because everything is rapidly passing. To use the display and the USB data connection, some terminal strips can be soldered making both work.

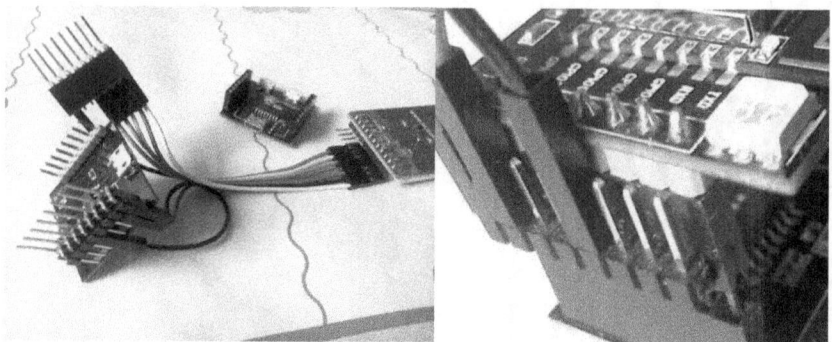

Figure 42: 90 ° headers provide a plug-in adapter for the ESP-module

To test the display, a bare minimum instead of the entire graphic demo will do. To print the milliseconds only as a graphic text, a sketch will look like this:

```
#include <SPI.h>
#include <Adafruit_GFX.h>     // Core graphics library
#include <Adafruit_ST7735.h> // Hardware-specific library

#define TFT_CS    15
#define TFT_RST   12
#define TFT_DC    16
#define TFT_SCLK 14
#define TFT_MOSI 13
Adafruit_ST7735 tft = Adafruit_ST7735(TFT_CS,TFT_DC,
TFT_RST);

void setup(void) {
  tft.initR(INITR_BLACKTAB);
  tft.setTextWrap(true);
  tft.fillScreen(ST7735_BLACK);
  tft.setTextColor(ST7735_WHITE,ST7735_BLACK);
  tft.setTextSize(3);
}
```

```
void loop()
{tft.setCursor(0, 0);
 tft.print(millis());
}
```

Figure 43: ESP8266 controlling a TFT display

1.2.9 *ESP8266-CORE: HOTSPOT/ACCESS-POINT*

In order not to reproduce too much source code here, the core examples should serve as a template with as few changes as possible. At *ESP8266WiFi file/examples* the most important are summarized.

To let the ESP operate as an own hotspot or access point (AP) there is a template WiFiAccessPoint available. In that code the name and the password is set to "ESPap" and "thereisnospoon", which of course is cus-tomizable. As the server library is involved, the connection can be checked using a browser on any device that connects to the hotspot. The server is listening to the standard port 80 and on connect an output is displayed in the function *handleRoot ()* "You are connected" as a header 1 in html.

Configuring access point...

AP IP address: 192.168.4.1

HTTP server started

The device to be connected now logs in by password, then the browser calls the address http://192.168.4.1/, whereupon the corresponding connection confirmation is displayed.

This is the way the ESP can be reached from any browser - even without an Internet infrastructure - if there is no 2.4 GHz jammer nearby.

1.2.10 ESP8266-CORE: INTERNET-ACCESS

An Internet connection is usually done by something like a router. As with a smart phone, the name of the router and the password must be known. Then an ESP connection is done by the following lines, which are adapted from the many templates.

```
#include <ESP8266WiFi.h>

bool Internet(char *ssid, char *pass);

void setup()
{Serial.begin(9600); delay(5000);
 Serial.println("Internettest in 5 Sekunden.");
 delay(5000);
 char ssid[] = "FRITZ!Box SL WLAN";   // your net-
work SSID (name)
 char pass[] = "xxxxxxxxxxxxx";        // your net-
work password
 Serial.println(Internet(ssid,pass));
}

bool Internet(char *ssid, char *pass)
{int i=0;
```

```
WiFi.disconnect();
WiFi.begin(ssid, pass);
Serial.print("\nVerbinde mit ");
Serial.print(ssid);
while (WiFi.status() != WL_CONNECTED && i++ < 20)
{delay(500);Serial.print(".");}
if(i == 21)
{Serial.println("\nKein Internet.");
 return false;
}
Serial.println("\nInternet ok.");
 return true;
}

void loop() {}
```

The router has to be set to accepting new devices. Outdoors a smartphone can be used as a router creating a mobile hotspot (tethering). If the connection is successful, the exact time may be requested.

1.2.11 ESP8266-CORE: INTERNET-TIME/CLOCK

The following slightly longer listing for the ESP is a combination of several sources. The result is the representation of an analogue clock on a TFT display, synchronized by a time server. The designs origin is at *http://www.hjberndt.de/soft/BTUHR.html*.

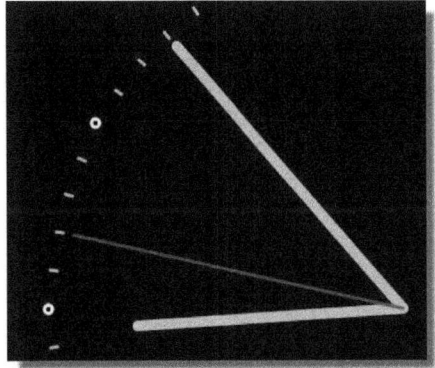

Figure 44: JavaScript-Clock (exerpt)

Since an ESP8266 now easily is connected to the Internet, it is the time to obtain the exact time from there. In the section ESP8266-AT this is done in an unusual way, because at the time of the project no such simple routines were available. The ESP8266-CORE provides a finished Sketch *NTPClient.ino* in the ESP8266WiFi examples. Only the name of the AP and its password is to be supplemented. Here the output via the serial monitor with 115,800 Baud:

sending NTP packet...
packet received, length=48
Seconds since Jan 1 1900 = 3686839151
Unix time = 1477850351
The UTC time is 17:59:11
sending NTP packet...
packet received, length=48
Seconds since Jan 1 1900 = 3686839162
Unix time = 1477850362
The UTC time is 17:59:22

The test program endlessly repeats the request and the output. Meanwhile, the *TimeLib* or *Time Library* has been expanded by a clock example for the ESP8266. With the same procedure concerning the AP name and password this sketch shows the following output – now using 9600 baud:

TimeNTP Example
Connecting to FRITZ!Box SL WLAN
.....IP number assigned by DHCP is 192.168.178.31
Starting UDP
Local port: 8888
waiting for sync
Transmit NTP Request
Receive NTP Response
19:10:14 30.10.2016
19:10:15 30.10.2016
19:10:16 30.10.2016
19:10:17 30.10.2016
19:10:18 30.10.2016
19:10:19 30.10.2016
...

The analogue clock has formerly been presented at the following link, when it came to get the exact time via GPS:

http://www.hjberndt.de/soft/ardgpstime.html

Further down the page there is a "watch cabinet" with four different watch faces. The clock faces for the TFT display can be found at the end of the skits. By combining the above examples for NTP Time and the old Arduino source code, the analogue clock for the ESP8266 on a TFT display with synchronization from the internet is created. Due to shortness of the listing daylight savings time is not included.

```
#include <TimeLib.h>
#include <ESP8266WiFi.h>
#include <WiFiUdp.h>
#include <SPI.h>
#include <Adafruit_GFX.h>
#include <Adafruit_ST7735.h>

#define TFT_CS    15
#define TFT_RST   12
#define TFT_DC    16
#define TFT_SCLK  14
#define TFT_MOSI  13
Adafruit_ST7735 tft = Adafruit_ST7735(TFT_CS,   TFT_DC,
TFT_RST);

//----- SOME  TFT-COLORS
#define rgb     tft.Color565
#define ORANGE rgb(255,191,0)
#define BLACK  rgb(0,0,0)
#define WHITE  rgb(255,255,255)
#define RED    rgb(255,0,0)
#define GRAY   rgb(192,192,192)
#define RADDEG 0.0174532925
#define min(a,b) (((a)<(b)) ? a : b)
char *Tag[7] =
{"SONNTAG","MONTAG","DIENSTAG","MITTWOCH","DONNERSTAG","F
REITAG","SONNABEND"};

int s10=0,s01=0;
int m10=0,m01=0;
int h10=2,h01=3;

#define TICKPIN  3
```

```
#define CW 7          //CHAR. W
#define CH 8

char ssid[] = "EasyBox-8E5B26";  //  your network SSID
(name)
char pass[] = "xxxxxxxxx";         // your network password

IPAddress timeServer(132, 163, 4, 101);
// IPAddress timeServer(132, 163, 4, 102);
// IPAddress timeServer(132, 163, 4, 103);

const int timeZone = 1; // Central European Time

WiFiUDP Udp;
unsigned int localPort = 8888;
void sendNTPpacket(IPAddress &address);
time_t getNtpTime();
void digitalClockDisplay();
void printDigits(int digits);
void drawClock(boolean hands);

void setup()
{tft.initR(INITR_BLACKTAB);
 tft.setTextWrap(true);
 tft.fillScreen(BLACK);
 tft.setTextColor(WHITE,BLACK);
 tft.setTextSize(1);
 drawClock(false);
 WiFi.disconnect();
 Serial.begin(9600);
 delay(250);
 Serial.println("TimeNTP Example");
 Serial.print("Connecting to ");
 Serial.println(ssid);
 WiFi.begin(ssid, pass);
 while (WiFi.status() != WL_CONNECTED)
 {delay(500);
  Serial.print(".");
 }
 Serial.print("IP number assigned by DHCP is ");
 Serial.println(WiFi.localIP());
 Serial.println("Starting UDP");
 Udp.begin(localPort);
 Serial.print("Local port: ");
 Serial.println(Udp.localPort());
 Serial.println("waiting for sync");
```

```
 setSyncProvider(getNtpTime);setSyncInterval(900);
}

void tftclock()
{char s[20];
 tft.setTextSize(2);tft.setCursor(0,0);
 sprintf(s,"%02d:%02d:%02d",hour(),minute(),second());
 tft.print(s);
}

void loop()
{static int os=-1,om=-1,oh=-1;
 int w=tft.width(); int h=tft.height();
 if(second()!=os)         //every second
 {s01=second() % 10; s10=second() / 10;
  m01=minute() % 10; m10=minute() / 10;
  h01=hour()    % 10; h10=hour()    / 10;
  drawClock(true);
  if(om!=minute())        //every minute
  {if(oh!=hour()) //every hour
   {tft.setCursor(2,h-10);
    tft.print(char('0'+h10));
    tft.print(char('0'+h01));
    tft.print(':');
    oh=hour();
   }
   tft.setCursor(3*CW-2,h-10);
   tft.print(char('0'+m10));
   tft.print(char('0'+m01));
   tft.print(':');
   //DATE every minute
   tft.setCursor(2,2);
   if(day()<=9)tft.print('0');
   tft.print(day());tft.print('.');
   if(month()<=9)tft.print('0');
   tft.print(month());tft.print('.');
   tft.print(String(year()).substring(2));
   om=minute();
  }
  //every second
  tft.setCursor(6*CW-6,h-CH-2);
  tft.print(char('0'+s10));
  tft.print(char('0'+s01));
  os=second();
 }
}
```

```
void digitalClockDisplay()
{Serial.print(hour());
 printDigits(minute());
 printDigits(second());
 Serial.print(" ");
 Serial.print(day());
 Serial.print(".");
 Serial.print(month());
 Serial.print(".");
 Serial.print(year());
 Serial.println();
}

void printDigits(int digits)
{Serial.print(":");
 if(digits < 10)
 Serial.print('0');
 Serial.print(digits);
}
/*-------- NTP code ----------*/
const int NTP_PACKET_SIZE = 48;
byte packetBuffer[NTP_PACKET_SIZE];

time_t getNtpTime()
{while (Udp.parsePacket() > 0) ;
 Serial.println("Transmit NTP Request");
 sendNTPpacket(timeServer);
 uint32_t beginWait = millis();
 while (millis() - beginWait < 1500)
 {int size = Udp.parsePacket();
  if (size >= NTP_PACKET_SIZE)
  {Serial.println("Receive NTP Response");
   Udp.read(packetBuffer, NTP_PACKET_SIZE);
   unsigned long secsSince1900;
  secsSince1900 =  (unsigned long)packetBuffer[40] << 24;
  secsSince1900 |= (unsigned long)packetBuffer[41] << 16;
  secsSince1900 |= (unsigned long)packetBuffer[42] << 8;
  secsSince1900 |= (unsigned long)packetBuffer[43];
   return secsSince1900 - 2208988800UL + timeZone *
SECS_PER_HOUR;
  }
 }
 Serial.println("No NTP Response :-(");
 return 0;
}
```

```
void sendNTPpacket(IPAddress &address)
{// set all bytes in the buffer to 0
 memset(packetBuffer, 0, NTP_PACKET_SIZE);
 // Initialize values needed to form NTP request
 // (see URL above for details on the packets)
 packetBuffer[0] = 0b11100011;   // LI, Version, Mode
 packetBuffer[1] = 0;      // Stratum, or type of clock
 packetBuffer[2] = 6;      // Polling Interval
 packetBuffer[3] = 0xEC;  // Peer Clock Precision
 // 8 bytes of zero for Root Delay & Root Dispersion
 packetBuffer[12]  = 49;
 packetBuffer[13]  = 0x4E;
 packetBuffer[14]  = 49;
 packetBuffer[15]  = 52;
 // all NTP fields have been given values, now
 // you can send a packet requesting a timestamp:
 Udp.beginPacket(address, 123); //NTP requests are to
port 123
 Udp.write(packetBuffer, NTP_PACKET_SIZE);
 Udp.endPacket();
}

// CLOCK ---------------------------------------------
// Gilchrist 6/2/2014 1.0
// Updated by Alan Senior 5/1/2015
// Modified for own needs: H.-J. Berndt 3/2015
// ---------------------------------------------------

void drawClock(boolean hands)
{static float sx = 0, sy = 1, mx = 1, my = 0, hx = -1, hy
= 0;
 static float sdeg=0, mdeg=0, hdeg=0;
 static uint16_t osx, osy, omx, omy, ohx, ohy;
 static uint16_t x0, x1, y0, y1;
 static boolean initial = 1;
 static int w,h,w2,h2;
 if(!hands)
 {w=tft.width();w2=w/2;
  h=tft.height();h2=h/2;
  osx=w2, osy=h2, omx=w2, omy=h2, ohx=w2, ohy=h2;
  tft.setTextColor(GRAY,BLACK);
  for(int i = 0; i<360; i+= 6)
  {sx = cos((i-90)*RADDEG);
   sy = sin((i-90)*RADDEG);
   x0 = w2+sx*(min(w2,h2)-4);
```

```
  y0 = h2+sy*(min(w2,h2)-4);
  if(i%30)tft.drawPixel(x0, y0, WHITE);
  else tft.drawCircle(x0,y0,1,WHITE);
 }
}
else
{sdeg = second()*6;
 mdeg = minute()*6;
 hdeg = (hour()% 12)*30 + mdeg/12;
 hx = cos((hdeg-90)*RADDEG);
 hy = sin((hdeg-90)*RADDEG);
 mx = cos((mdeg-90)*RADDEG);
 my = sin((mdeg-90)*RADDEG);
 sx = cos((sdeg-90)*RADDEG);
 sy = sin((sdeg-90)*RADDEG);
 if (second()==0 || initial)
 {initial = 0;
  tft.drawLine(ohx, ohy, w2, h2, BLACK);
  tft.drawLine(ohx-1, ohy-1, w2-1, h2-1, BLACK);
  tft.drawLine(ohx+1, ohy+1, w2+1, h2+1, BLACK);
  ohx = w2+hx*(min(h2,w2)-20);
  ohy = h2+hy*(min(h2,w2)-20);
  tft.drawLine(omx, omy, w2, h2, BLACK);
  omx = w2+mx*(min(h2,w2)-10);
  omy = h2+my*(min(h2,w2)-10);
 }
 tft.drawLine(osx, osy, w2, h2, BLACK);
 tft.drawLine(ohx, ohy, w2, h2, ORANGE);
 tft.drawLine(omx, omy, w2, h2, ORANGE);
 osx = w2+sx*(min(h2,w2)-8);
 osy = h2+sy*(min(h2,w2)-8);
 tft.drawLine(osx, osy, w2, h2, RED);
 tft.fillCircle(w2, h2, 2, RED);
 }
}
```

1.2.12 ESP8266-CORE: SERIAL-WI-FI-CONVERTER (GPS)

The ESP9266 is often called a serial to Wi-Fi converter. In the beginning control and programming was distributed only with AT sequences via its serial interface. With the possibility of using the IDE Core for coding and control, this serial interface is now available for this kind of conversion. To illustrate this 9600 baud data stream of a GPS module is to be disseminated unchanged over Wi-Fi.

Figure 45: GPS to Wi-Fi hardware

Using the core template *WiFiTelnetToSerial* it is possible without changes. The sketch represents a server which is capable of contact with up to 3 clients through port 333 and passes the data to the serial interface. The specified alternative listing is a shortened modification and uses its own AP named "ESPap".

```
#include <ESP8266WiFi.h>

WiFiServer server(333);
WiFiClient serverClients[3];

void setup()
{Serial.begin(9600);
 WiFi.disconnect();
 delay(100);
 WiFi.softAP("ESPap", "");
 server.begin();
 server.setNoDelay(true);
 Serial.print(WiFi.softAPIP());
 Serial.println(":333 to connect");
}

void loop()
{uint8_t i;
 // CLIENTS
 if(server.hasClient())
 {for(i = 0; i < MAX_SRV_CLIENTS; i++)
  {if (!serverClients[i] || !serverCli-
ents[i].connected())
```

```
 {if(serverClients[i])serverClients[i].stop();
  serverClients[i] = server.available();
  Serial1.print("New client: "); Serial1.print(i);
  continue;
  }
 }
 WiFiClient serverClient = server.available();
 serverClient.stop();
}
// COPY TO SERIAL
for(i = 0; i < MAX_SRV_CLIENTS; i++)
{if (serverClients[i] && serverClients[i].connected())
 {if(serverClients[i].available())
  {while(serverClients[i].available())
   {char c = serverClients[i].read();
    Serial.write(c);
   }
  }
 }
}
// COPY FROM SERIAL
if(Serial.available())
{size_t len = Serial.available();
 uint8_t sbuf[len];
 Serial.readBytes(sbuf, len);
 for(i = 0; i < MAX_SRV_CLIENTS; i++)
 {if (serverClients[i] && serverClients[i].connected())
  {serverClients[i].write(sbuf, len);
   delay(1);
  }
 }
 }
}
```

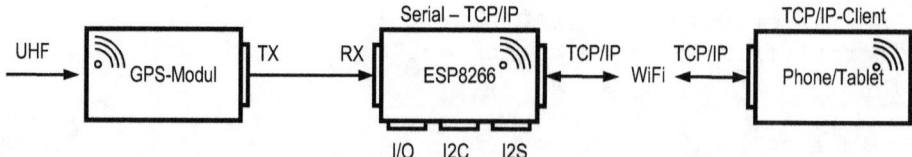

Figure 46: GPS Wi-Fi and a receiver example

All outputs and inputs from and to the serial interface are now redirected to port 333 of the IP. The main loop checks the client connections and the two transmission channels. To check this Sertial2Wifi connection on a smartphone, the following steps are advised:

- Connect mobile device to the AP „ESPap".
- In a TCP/IP client, call the IP 192.168.4.1 on port 333

Both the *Wi-Fi TCP Test Tool* running on the iPod, the *TCP Client* on Android, *rfo*-Basic on Android and of course HyperTerminal or *RealTerm* on Windows display the text data stream.

1.2.13 ESP8266-CORE: CONTROL USING WI-FI

A relay is switched in the section ESP8266-AT via Wi-Fi. Contol using Wifi in this book and as a supplement to [2] is available at:

http://www.hjberndt.de/soft/ardesp8266.html

This now can be done without an Arduino as follows.

The many examples or templates of the ESP-core IDE allow easy conversion in a few steps. In fact the sketch *WiFiTelnetToSerial* needs to be extended at only one place. In the section containing the comment *// get data from the telnet client and push it to the UART*, the following lines needs to be added:

```
//get data from the telnet client and push it to the UART
while(serverClients[i].available())
{char c = serverClients[i].read();
 if(c=='1'){digitalWrite(13, HIGH); serverCli-
ents[i].println("An");}
 if(c=='0'){digitalWrite(13, LOW); serverCli-
ents[i].println("Aus");}
}
```

In the upper part, the placeholders for router name and password, as in almost all examples, has to be correspondingly added:

```
const char* ssid = "**********";
const char* password = "**********";
```

The digital output 13 of the ESP8266/12F and its blue LED now controls a relay, according to the incoming signal, as in the original, but now an Arduino is no longer needed.

1.2.14 ESP8266-CORE: SIMPLE-BASIC

A sketch of 1717 lines in a single *.ino file is the original version of *ESPBasic* from September 2015. This relatively short sketch by Michael Molinari aka *mmiscool* is the minimum, which later in a version 2 and most recently in 2016 in a revised version 3 (Branch) is published on *https://www.esp8266basic.com*. The section ESPBasic in this book uses primarily version 3. All three variants are open source and are available at github. The second version divides the source code for overview reasons and scope into multiple files, so an entire folder of source code has to be compiled and after some forward declarations this works fine using the current IDE.

https://github.com/mmiscool/Basic/blob/master/ESP8266Basic/ESP8266Basic.ino

However, version 1 is kind of charming, showing what it is in line 151 of the source code on github at the given url above:

```
151 PrintAndWebOut("Simple BASIC Interperter For
ESP8266...");
```

Maybe development began with the example Web server and the sspiff-filesystem of the core on which the sketch relies on. The instruction set is gratifying clearly and for many tasks quite adequately with the advantage that the BASIC lives with his applications in the ESP and is available or programmed wireless via browser. After a short source code analysis, the point of interpreting the commands is found and there is the place to let explode your own creativity.

1.2.15 ESP8266-CORE: NEW BASIC-COMMAND

It is astonishing that an important command is not implemented in version 1. The "Hello World" of the LED blink is to be realized here "cumbersome" using the goto command:

[start]
po 13, 1
delay 500
po 13, 0
delay 500

goto [start]

Wouldn't it be nice to do a *Blink 13, 10, 500, 500* using one command, wherein the four parameters are the pin, the repetitions and the on- and off delay. This command should be inserted within the routine *ExicuteTheCurrentLine()* from the 715th source code line following the existing command *po*, as used above:

```
if (Param0 == "blink")   {
  valParam1 = GetMeThatVar(Param1).toInt();
  valParam2 = GetMeThatVar(Param2).toInt();
  valParam3 = GetMeThatVar(Param3).toInt();
  valParam4 = GetMeThatVar(Param4).toInt();
  pinMode(valParam1, OUTPUT);
  for(int i = 0; i < valParam2; i++)
  {digitalWrite(valParam1,HIGH);delay(valParam3);
   digitalWrite(valParam1,LOW );delay(valParam4);
  }
  return;
}
```

After uploading the modified Basic-Sketch, a 10-times flashing of the LED on pin 13 with a frequency of 1 Hz and pulse duration of 100 ms, as well as a pulse pause of 900 ms can now be called in one line:

blink 13 10 100 900

Using variables (no commas as a parameter separation) works fine too. If at compile time error messages arise of the type "xyz what not declared in this scope", the IDE compiler expects forward declarations. These declarations listed down here can be inserted at line 37.

```
void PrintAndWebOut(String itemToBePrinted);
void SetMeThatVar(String VariableNameToFind, String NewContents);
String evaluate(String expr);
String VarialbeLookup(String VariableNameToFind);
String RunningProgramGui();
String GetRidOfurlCharacters(String urlChars);
String FetchWebUrl(String URLtoGet);
String DoMathForMe(String cc, String f, String dd );
String GetMeThatVar(String VariableNameToFind);
String FetchWebUrl(String URLtoGet);
```

```
void CreateAP(String NetworkName, String NetworkPassword);
byte ConnectToTheWIFI(String NetworkName, String NetworkPass-
word);
String LoadDataFromFile(String fileNameForSave);
void SaveDataToFile(String fileNameForSave, String DataToSave);
void LoadBasicProgramFromFlash(String fileNameForSave);
String getValueforPrograming(String data, char separator, int
index);
void SaveBasicProgramToFlash(String fileNameForSave);
byte CheckFOrWebGOTO();
void CheckFOrWebVarInput();
int RunBasicTillWait();
void ExicuteTheCurrentLine();
void PrintAllMyVars();
```

1.3   ESP8266-AT (using Arduino)

At the time the *ESP8266* was released, the first applications in the hobby area were created using AT commands. Even old telephone modems from the early days of the Internet used this type of communication, but even today the automobile OBD interface uses this type for its on-board computer.

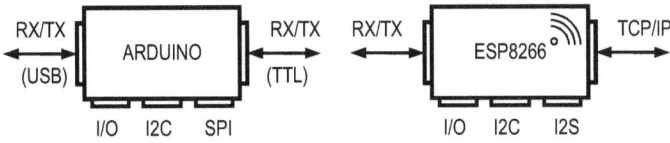

Figure 47: Tandem: Arduino and ESP8266

To get things going an Arduino sent AT commands in the desired order. First experiments concerning "Control with Wi-Fi" were done and short reference can or could be found at:

https://github.com/espressif/ESP8266_AT/wiki.

1.3.1   ESP8266-AT: Commands in Brief

General
AT Test
AT + RST Restart
AT + GMR Version query
ATE Echo on / off

Wi-Fi
AT + CWMODE Wi-Fi mode
AT + CWJAP Connect to AP
AT + CWLAP List of received AP
AT + CWQAP Terminate connection with AP
AT + CWSAP Set parameters in AP mode
AT + CWLIF Show connected IPs

AT + CWDHCP DCHP on/off
AT + CIPSTAMAC Set mac address for Station
AT + CIPAPMAC Set mac address for AP
AP AT + CIPSTA Set IP for Station
AT + CIPAP Inquire IP of AP

TCP
AT + CIPSTATUS, Inquire connection status
AT + CIPSTART beginning TCP/UDP connection
AT + CIPSEND transmission data
AT + CIPCLOSE Complete compound
AT + CIFSR, Inquire local IP
AT + CIPMUX multiple mode
AT + CIPSERVER configuration as the server
AT + CIPMODE transmission mode
AT + CIPSTO Set timeout for server
AT + CIUPDATE update via network
+ IPD Receive data

On the site, there are some examples that can be tested with a terminal program without Arduino. Using terminal software the *ESP8266* can be tested without the upload of own code. Since there still may be applications for those now cumbersome procedures, here are some own written examples on the principle of AT commands. In this way, the device can be controlled directly via its serial interface. The current version 1 of this serial to Wi-Fi converter uses at default 9600 baud and works on his RX/TX lines with a level of 3.3 volts The first steps can be done using standard terminal software. The level adaptation of TTL (5 V) to 3.3 V can be carried out via an appropriate level converter, a FTDI-Adapter with 3.3 V supply, a 3.3 V Arduino or ignoring this fact at your own risk and still using 5 Volts. At least the version 1 of the ESP has survived here doing so until now.

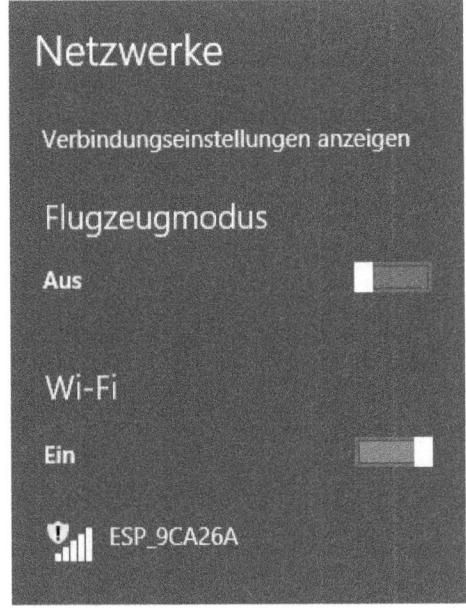

Figure 48: ESP as AP

1.3.2 ESP8266-AT: MANUAL CONTROL

Due to the availability of an *FTDI-adapter* providing 3.3 V this hardware is used from the various available options. A jumper has to be plugged only. The connection of the adapter is then as follows:

If V_{cc} is Pin 1 (upper right) and pin 8 Gnd (bottom left), the connections following a line-by-line numbering:

ESP		FTDI
1	Vcc	3.3 V
2	RX	TX
3	–	–
4	–	–
5	ChP	3.3 V
6	–	–
7	TX	RX
8	GND	GND

Figure 49: Connections cf. Fig. below

In case of issues because of the Wi-Fi transmit power with this compound, a capacitor of several microfarads between ground and 3.3 Volts remedy the situation. The module handles the serial communication only as a background task, its main purpose is Wi-Fi, and the load of the 3.3 Volt powersource is unknown. HyperTerminal is being used in the following testing as it is an excellent tool for this experiment. Each transmitted AT command must end with a newline. The ASCII setup dialog allows to set the hook at the appropriate place.

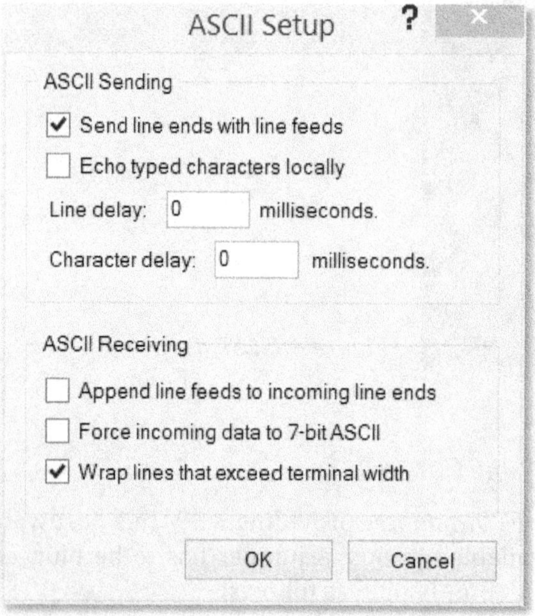

Figure 50: Hyperterminal: ASCII-Setup

If by mistake the echo enabled in the dialog above, all entries appear (here "AT") twice, as the ESP produces itself an echo in the default setting.

```
AAATT

OK
```

Next there are some manual commands. The first step is to perform a reset with the sequence "AT + RST" after a while:

```
AT+RST

OK
ê
[Vendor:www.ai-thinker.com Version:0.9.2.4]
```

```
ready
```

The version of the operating system is shown using "AT + GMR"

```
AT+GMR
0018000902-AI03

OK
```

Using "AT + CWMODE", the mode is set or requested. A missing question mark in the query will be answered accordingly.

```
AT+CWMODE
no this fun
AT+CWMODE?
+CWMODE:3

OK
```

To list the received hotspots in the reception area is entering "AT + CWLAP".

```
AT+CWLAP
+CWLAP:(0,"dlink",-88,"00:26:5a:b1:cb:52",9)
+CWLAP:(3,"TeliaGateway9C-97-26-B4-AD-DB",-
83,"9c:97:26:b4:ad:db",11)
+CWLAP:(3,"TeliaGatewayA4-B1-E9-BD-DF-13",-
95,"a4:b1:e9:bd:df:13",11)

OK
```

Three access points are available and even an open hotspot "dlink" with - 88 dB and its MAC on channel 9. As shown above, the attached ESP is working as a separate access point or hotspot because this mode has been set elsewhere.

The IP of this AP ESP is gotten by "AT + CIFSR", but without question marks.

```
AT+CIFSR?
no this fun
```

```
AT+CIFSR
192.168.4.1
0.0.0.0

OK
```

At the time of interrogation there was no local router connection, so this IP is 0.0.0.0. The short session as a screenshot:

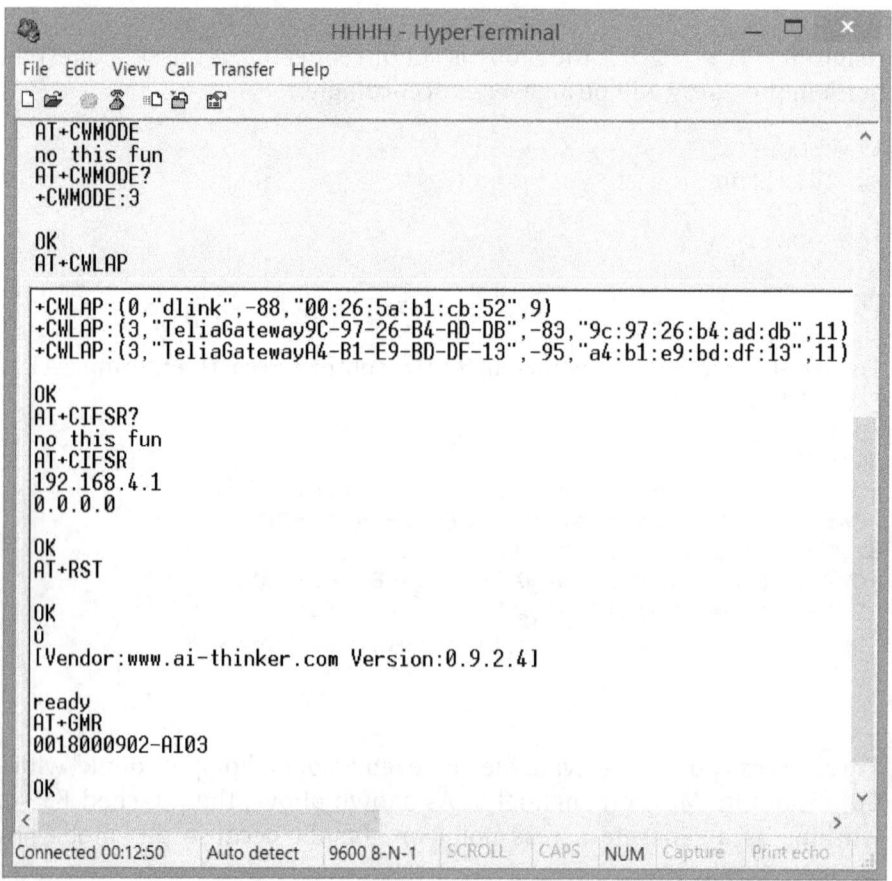

Figure 51: Manual AT control commands to ESP8266-1 in HyperTerminal

Doing like this it is possible to initiate the connection with a router once. Since password and SSID remain stored in the ESP, a new registration is not required. The following statements are based on this approach, so do not show up these two strings in the lines. Just putting the ESP in that mode using the AT command will do.

1.3.3 *ESP8266-AT: ARDUINO AS MEDIATOR*

The manual AT control can run automatically using a microcontroller and serial tranfer. So it would be possible to use the Digispark, however, an Arduino offers, as in the development of the upload periods shorter or giving more convenient control via serial monitor. To use both the serial monitor and the ESP, the Arduino requires two serial interfaces. With the library *SoftwareSerial* in addition to standard connections pins 0/1 two more pins 2/3 can be set up as an interface for the ESP. This second RX/TX connection is called esp below. If the test phase ended successfully, only one serial connection is needed.

```
#include <SoftwareSerial.h>
#define DEBUG true
#define PIN 12
SoftwareSerial esp(2,3); // (ESP1-Pin 7/2)
```

Now it should be possible to send AT commands using *esp.println*. The corresponding routine could look like this:

```
void espSendLine(char *s)
{esp.println(s);
}
```

In a terminal program the answer immediately appears in a window, but in code a small reading routine must ensure that the expenditure of ESP arrive. Since some commands may need a while to e.g. the surrounding hotspots scanned, the read routine takes a delay in milliseconds as a second parameter. A 0 skips this delay.

```
char *espReadLine(int ms)
{char s[200]="";
 if(ms)delay(ms);
 int len=esp.readBytesUntil('\n',s,sizeof(s)-8);
s[len]=0;
 if(len)if(DEBUG)Serial.println(s);
 return s;
}
```

Since all the answers end up with a newline, everything up to this character '\n' is read into the buffer *s* and finally returned null-terminated. If
DEBUG is true, the reply appears in the serial monitor of the IDE.

Using these two routines the desired command sequences can now be
coded. For the ESP, some kind of initialization is required, which could
look like this:

```
int espInit()
{int stat=88;
 esp.println("ATE0");              espReadLine(0);
 esp.println("AT+CWMODE=2");       espReadLine(0);
 esp.println("AT+CIFSR");          espReadLine(0);
 esp.println("AT+CIPMUX=1");       espReadLine(0);
 esp.println("AT+CIPSERVER=1");    espReadLine(0);
 esp.flush();
 esp.println("AT+CIPSTATUS");
 sscanf(espReadLine(0),"STATUS:%d",&stat);
 if(DEBUG)Serial.println(stat);
 return stat;
}
```

First, the echo is turned off, since this is more of a hindrance at 9600
baud and in this environment. With CWMODE's the own AP is turned on,
CIFSR reports the local IP. CIPMUX and CIPSERVER set the server mode
to multiple-mode. At the end CIPSTATUS reports the connection status.
Now the ESP is initialized accordingly. Next is the initialization of the
Arduino itself. The *setup()* initializes the terminals of the two interfaces,
as well as pins 12 and 13 as the output of an LED.

```
void setup()
{pinMode(PIN,OUTPUT);  pinMode(13,OUTPUT);
 Serial.begin(9600);   esp.begin(9600);
 Serial.println("ESP8266 LED SWITCH");
 digitalWrite(13,HIGH); //BUSY
 espInit();
 digitalWrite(13,LOW);//READY
}
```

In this first version a LED will be controlled using TCP/IP. This sketch is
the precursor of the next section, which was found on the web as "Control using Wi-Fi", Google can find it by searching the original "Steuern

mit WLAN".

An Arduino needs a main loop next to initialization. There the incoming character "0" results in a LOW-output and the corresponding HIGH for the LED at pin 12 receiving a "1".

```
void loop()
{int id,len,i=0;char s[80];
 if(3==sscanf(espReadLine(0),"+IPD,%d,%d:%s",id,len,s))
 {digitalWrite(PIN,s[0]=='0'?LOW:HIGH); // OUTPUT
  sprintf(s,"AT+CIPSEND=%d,10\r\n",id);
  esp.print(s);espReadLine(0);
  if(digitalRead(PIN))esp.print("LED is ON ");
  else esp.print("LED is OFF");
 }
}
```

The ESP routes the sent data via TCP/IP to the serial interface in a special format. In addition to the message, the ID of the sender and the message length is transmitted. If the first character is a "0", the LED is turned off, otherwise turned on. Finally, it is reported back as the state of the LED is in such a way that the message is 10 characters long, since the ESP expects this information when using CIPSEND. Substituting the above source lines together in this order, a compliable Arduino sketch should become available, which works exactly as the in advance published web-variant, which is formulated even more compact in some places, but gives the same result.

After start and a connected LED to pin 12 of the UNO, the smartphone will search for an AP named "ESPxxxx" and connects to a TCP/IP client 192.168.4.1 on port 333. If there is no hotspot of the ESP in the air, this could be done by AT commands yet. This setting is stored, so there is no need to activate the AP every time. Tapping on the smartphone a 1 or a 0, the LED should respond at pin 12. In the Serial Monitor following output might show up:

```
ESP8266 LED SWITCH
O_ÿ
no change
192.168.4.1
-5þ
O5ÿno Ð¹•RSTATUS:2
```

```
88
OK
Link
+IPD,0,2:1
OK
>
SEND OK
+IPD,0,2:0
OK
>
SEND OK
Unlink
```

After an Arduino reset, a re-initialization of the ESP and an already bandaged smartphone, one or more lines might possibly appear after the 88 as

+CIPSTATUS:0,"TCP","192.168.4.100",43320,1.

As a complement to ebook [2] other methods of interaction are given in the following sections. There, in 2015, the topic TCP/IP and WLAN was excluded at the time because it was foreseeable that the ESP through its popularity would get an even easier access, which in the form of the Arduino IDE - programming as Arduino, but without an Arduino - got true. The chapter *ESP8266*-Core uses the *ESP8266* like this.

1.3.4 ESP8266-AT: CONTROL USING WI-FI

The next application uses an Arduino as an intermediary and as a measuring and control unit. The example shown here does a relay control as a high voltage switch. This control is to take place via mobile devices such as smartphones or tablets, which can also take control simultaneously.

The following devices and components are required for this configuration:

- Arduino IDE (development environment)
- ESP8266-1 at 9600 baud
- A capacitor approx. 470 µF - 1000 µF
- A relay breakout (or only one LED)

- Jumper wire
- A smartphone, tablet or even a PC as a control unit

Figure 52:
Relay-control by Wi-Fi

Without the control unit, the cost of the hardware can remain in the single-digit range, an Arduino Mini included.

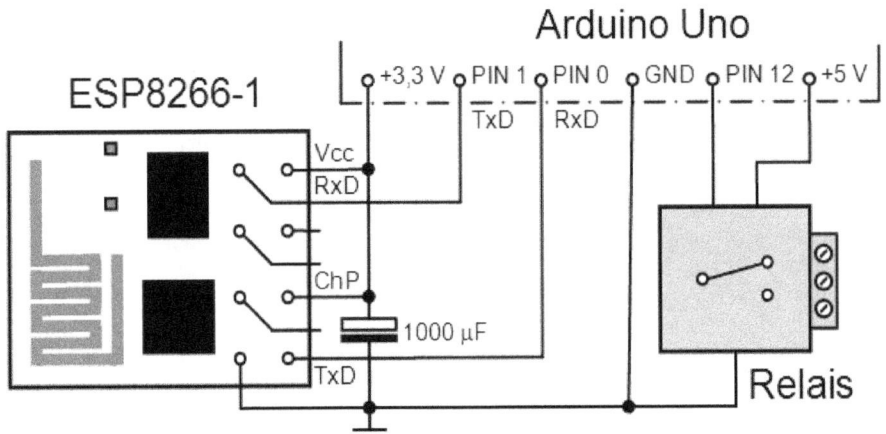

Figure 53: Connecting is done like this. Some caution concerning the power supply because the ESP8266 is sensitive to wrong tension. A capacitor stabilizes the weak 3.3 volts while transmitting. The direct connection of the TX/RX lines at your own risk

The Arduino runs a short sketch. It is a minimized version which does not require software Serial and is smoothly running after some Internet research. To upload the sketch lines 0/1 (RX/TX) must be disconnected from the *ESP8266*. After uploading, the connection is required because the communication between ESP and Arduino is done by these two standard serial pins. A PC is no longer needed now. In [2] is shown how programming and upload can be done from a smartphone - without a PC. The sketch initializes the serial at 9600 baud, the *ESP8266* as AP (CWMODE), which allows several clients and switches to server mode. If after a while issues arise, the ESP8266 resets (RST + AT) when A0 at an Arduino analogue input is at ground level. With *espReadLine()* characters are read from the *ESP8266*, the output is done by *Serial.print* or *Serial.println*. The main loop checks whether the character ("1") is received for switching on the relay.

A string coming from the ESP has the format "+ IPD, 2.5: Hello", if a user with ID 2 sends the 5 letter word Hello. If 3 parameters are detected, the test on "1" is done and the relay pin 12 is switched accordingly. LED is OFF or LED is ON. (This was done elsewhere in the Bluetooth control here as well). With CIPSEND the addressee (id) and the number (12) is transmitted to send the characters in advance and then the real message. Here, the current state of the pin 12 is polled.

```
//OHNE SOFT SERIAL
#define PIN 12
char *startup[]={"ATE0","AT+CWMODE=2","AT+CIPMUX=1",
                "AT+CIPSERVER=1","AT+CIPSTATUS"};
void setup()
{pinMode(PIN,OUTPUT); Serial.begin(9600);
 if(analogRead(A0)==0)
 {Seri-
al.println("AT+RST");delay(2000);espReadLine(1000);}
 for(int i=0;i<5;i++)
 {Serial.println(startup[i]);espReadLine(0);}
}

char *espReadLine(int ms)
{char s[80]=""; //buffer
 if(ms)delay(ms);
 int len=Serial.readBytesUntil('\n',s,sizeof(s)-4);
s[len]=0;
 return s;
}

void loop()
{int id,len;char s[80];
 if(3==sscanf(espReadLine(0),"+IPD,%d,%d:%s",&id,&len,s))
 {digitalWrite(PIN,s[0]== '1'? HIGH : LOW); // OUTPUT
  sprintf(s,"AT+CIPSEND=%d,12",id);// Message to id
```

```
Serial.println(s);espReadLine(0);// +\r\n
if(digitalRead(PIN))Serial.println("LED is ON ");
             else Serial.println("LED is OFF");
}
}
```

Especially with the Android TCP/UDP Commander, a Play Store app, the assembly controls without any issues. But even a small *rfo*-Basic program can switch the relay by touch. Even the old HyperTerminal can handle TCP/IP connections on Windows 8.1. At the iTunes store you will find similar applications.

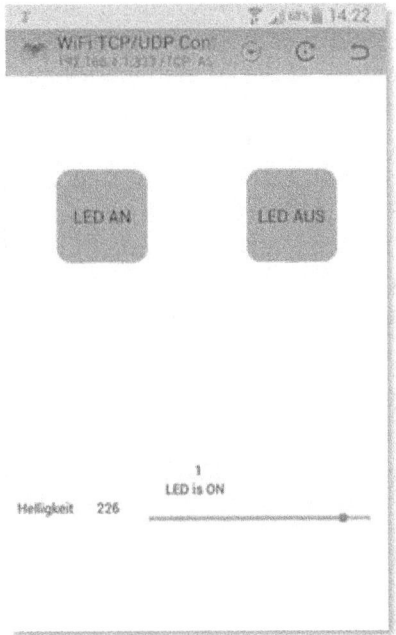

Figure 54: Control using PlayStore app: WiFi TCP/UDP Commander

The default IP of the ESP is 192.168.4.1 on port 333. This has been maintained to change things as less as possible. To get the control working on the smartphone or tablet the AP has to be changed. In the wireless settings an AP named "ESP _..." should be seen to connect. After that, a "Connect" made in in the adjacent app and pressing the button "LED ON" or sending a "1" in the hidden terminal window, the relay should switch. The "0" or "LED OFF" button switches off again. The commands are set in the app's settings. There are also slides, to send appropriate values.

Figure 55:
Touchcontrol/display (yel-
low circle) for LED-control

Sockets and TCP are supported by the easy to use rfo-Basic and so only the connection has to be changed in the old Bluetooth solution. This is done by "192.168.4.1" on port 333 via sockets. If the connection is established, the IP is issued, then after three seconds the graphics mode is used. On each touch the yellow circle is changed while a "1" or "0" is sent via Wifi to the *ESP8266*. This passes these characters to the two serial lines of an Arduino, which switches the relay accordingly. The pause in the touch query saves battery power of the smartphone. There still is an app available for the iPod Touch 2nd Generation in the App Store: *WIFI TCP Test Tool*. It is useful to examine the interaction of various apps and programs.

```
PRINT "WLan Steuerung mit ESP8266."
PRINT "Arduino Led13 schalten."
SOCKET.CLIENT.CONNECT "192.168.4.1",333
SOCKET.CLIENT.STATUS r
IF r THEN
 PRINT "Connected to ";
 SOCKET.CLIENT.SERVER.IP a$
 PRINT a$
ELSE
 END
ENDIF
PAUSE 3000

GR.OPEN 255, 0,0,0
GR.ORIENTATION 1
GR.SCREEN w,h
r = 255
g = 200
b = 0
GR.COLOR 255,r,g,b,1
```

```
GR.CIRCLE rc, w/2, h/2, w/3
lf$=CHR$(13)

DO
 GR.SHOW rc
 GR.RENDER
 SOCKET.CLIENT.WRITE.LINE "1"+lf$
 GOSUB touch
 GR.HIDE rc
 GR.RENDER
 SOCKET.CLIENT.WRITE.LINE "0"+lf$
 GOSUB touch
UNTIL x<100
END

touch:
DO
 GR.TOUCH touched,x,y
 PAUSE 20
UNTIL touched
WHILE touched
 GR.TOUCH touched,x,y
 !pause 20
REPEAT
RETURN
```

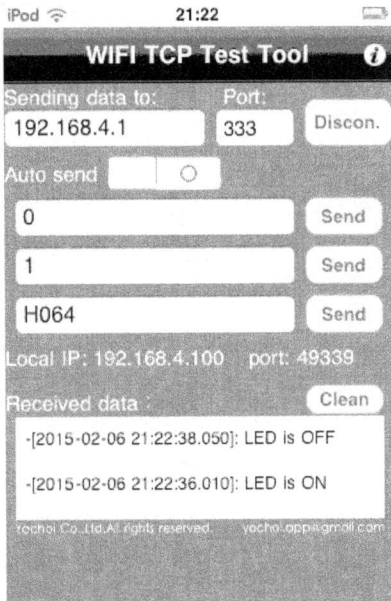

Figure 56: iPod Touch control via Wi-Fi

1.3.5 ESP8266-AT: Brightness Control

By doing a few additions, it is possible to implement analogue controls using PWM. In this example, a LED at pin 11 is intended to change its brightness as a function of a slider on a smartphone. To get this going, the Android *TCP/UDP Commander* app sends the letter "H" followed by a three-digit number in the range 000 to 255 for brightness. The brightness value H127 sends the APP via the ESP-Hotspot to the serial lines that are read by an Arduino. Again, the ESP is directly connected to Pin 0/1, so software Serial is not needed. In the main loop three initial characters now are distinguished and the next three digits following an "H" are converted to a decimal value to output as an analogue value to pin 11 on an Arduino. The rest of the Sketch predominantly corresponds to the preceding.

```
// ------------------------------------------------
// ESP8266 - TCP/UDP-Commander Android
// H000 - H255 wird ausgewertet, um
// Helligkeit einer LED zu steuern.
// Bezug: http://hjberndt.de/soft/ardesp8266.html
// ------------------------------------------------

#define LED 13 // Relais/Relay
#define SLD 11 // Helligkeit/Brightness

char *startup[]={"ATE0","AT+CWMODE=2","AT+CIPMUX=1",
                "AT+CIPSERVER=1","AT+CIPSTATUS"};
void setup()
{pinMode(LED,OUTPUT); Serial.begin(9600);
 if(analogRead(A0)==0)
 {Seri-
al.println("AT+RST");delay(2000);espReadLine(1000);}
 for(int i=0;i<5;i++)
 {Serial.println(startup[i]);espReadLine(0);}
}

char *espReadLine(int ms)
{char s[80]=""; //buffer
 if(ms)delay(ms);
 int len=Serial.readBytesUntil('\n',s,sizeof(s)-4);
s[len]=0;
 return s;
}
```

```
void loop()
{int id,len;char t[10],s[80];
 if(3==sscanf(espReadLine(0),"+IPD,%d,%d:%s",&id,&len,s))
 {switch (s[0])
   {case '0':digitalWrite(LED,LOW);break;
    case '1':digitalWrite(LED,HIGH);break;
    case 'H':analogWrite (SLD,(atoi(strncpy(t,&s[1],3))));
   }
   sprintf(s,"AT+CIPSEND=%d,11",id);// Message to id
   Serial.println(s);espReadLine(0);
   if(digitalRead(LED))Serial.print("LED is ON \r");
                 else Serial.print("LED is OFF\r");
 }
}
```

1.3.6 ESP8266-AT: CONTROL USING INTERNET - IOT

Until now, the ESP8266-1 itself served as an access point and thus could do without any further Wi-Fi infrastructure. In another mode, the ESP connects to a router and receives a local IP as each device to get internet access and this is how you get into the world of IoT (Internet of Things). The following example switches or controls something in a distant location.

At that location a wireless router is available and permanently online. The ESP has been associated manually once to this box by name and password. The BASIC structure is the same as used above in the relay control. However, such control must fulfill some conditions in order to work maintenance free. To get this done, some endurance tests were needed that finally ended successfully with the version 1 of the ESP8266 and the operating system version 0018000902-AI03 for a possible private application.

The external IPv4 here still changes every night, so a dynamic IP must do. At the time this Sketch was created the *TCP Commander* accepted numeric IP addresses only (the author of *TCP Commander* updated quickly to demand and added this feature), but many other apps accept even an URL like *EasyTCP* from the Play Store. This is how controlling the things on the internet can be done virtually via chat. In the example however, only the "1", is evaluated, the rest is pure gimmick, but extensions are possible. Rfo BASIC for Android supports URL names in his call

SOCKET.CLIENT.CONNECT too.

By presence of experimental and unshielded constructions, errors occur now and then. During the night an uncoupled Bluetooth device e.g. could interfere, which ends up in an uncontrollable state. To avoid this and other issues in continuous operation, the sledgehammer method is used practically, or suspenders and belts. A timer ensures that periodically a reset for the esp8266 occurs. To avoid the ESP fall into "deep sleep" automatically, in addition a periodic 'chat' is held. In detail, the code can look like this:

Figure 57 Chat using ESP

```
unsigned long t0,t1,tg;

void espReset()
{if(analogRead(A0)==0 || millis()>(t1+RRTIME))
{Serial.println("AT+RST");espReadLine(3000);t1=millis();}
  for(int i=0;i<5;i++)
  {Serial.println(startup[i]);espReadLine(0);}
```

```
}
```

The three timer variables *t0*, *t1* and *tg* are declared for these tasks. In setup, they are initialized, the time intervals themselves are defined as constants:

```
#define RTIME    180000
//TRIGGER ESP every RTIME   ms (3 min)
#define RRTIME  900000
//RESTART ESP every RRTIME ms (15 min)
#define GOTIME   (8*RRTIME)
//GOIP-Refresh 2 hour interval

void setup()
{pinMode(LED,OUTPUT); Serial.begin(9600);
 t0=millis();t1=t0;tg=t0;
 espReset();
}
```

The actual timer is only a query at the end of the main loop of whether the corresponding time has elapsed:

```
void loop()
{
 ...
 if(millis()>(t0+RTIME)){espReset();t0=millis();}
 ...
}
```

After 3 minutes an initialization is done without cutting the connection. Every 15 minutes in addition a reset is initiated, which is to ensure correct operation in between. For the test application that gave 'tolerable' short sporadic blockings or errors.

1.3.7 ESP8266-AT: DYNAMIC IP

Usually a fixed IP is required to be contstantly accessible by the same URL. As the providers usually distribute a new IP in the private sector here at night, the use of 'yesterdays' IP today is useless.

Many routers offer solutions to solve this problem at the time reconnecting. However, the existing router refuses to do the job using 'goip.de' as expected. For this and other reasons an independent method appears practical, this should be done by means of the ESP8266 itself. Updating with goip.de is done by the following call in the browsers address field:

http://www.goip.de/setip? Username = myurl & password = mypassword

The URL is myurl.goip.de and a fixed IP is then no longer needed as the dynamic IP is assigned to the URL by this service. To be globally accessible, the port in the router has to be enabled. In this case it is port 333 for the ESPs local IP as a device, 192.xxx.xxx.26 in the local network (AT + CIFSR). Because the call may not occur too often, an interval of 2 hours is fixed, which may result in a not (remote) controlable device at night for about 120 minutes. The query of the external IP is possible using AT commands and could be compared with the old IP to minimize goip calls.

Here is the part of the sketch to update in 2-hour intervals:

```
#define URL "goip.de"
#define DOMAIN "GET
http://www.goip.de/setip?username=meinname&password=...\r
\n"

void goip()
{char s[80];
 String cmd="AT+CIPCLOSE=4";// disconnect #4 if con
 Serial.println(cmd);espReadLine(0);
 cmd="AT+CIPSTART=4,\"TCP\",\"";cmd+=URL;cmd+="\",80";
 Serial.println(cmd);espReadLine(1000);
 cmd=DOMAIN;sprintf(s,"AT+CIPSEND=4,%d",cmd.length());
 Serial.println(s);   espReadLine(1000);
 Serial.print(cmd);   espReadLine(1000);
}
```

In response, the ESP8266 receives is something like this:

```
* * * * * * * * * * * * * * * * * * * * * * * * * * * * * * * * * * * * * * * * * * * * * * * * * * * * * * *
GoIP.de Updater
* * * * * * * * * * * * * * * * * * * * * * * * * * * * * * * * * * * * * * * * * * * * * * * * * * * * * * *

Aktualisierung wurde erfolgreich durchgeführt.

EOT
```

In this case, possible reception errors are ignored, only the proper inquiry is important at *goip.de*. This method could be successfully with other dynamic services using corresponding calls.

A final check was carried out for this structure at a distance of 200 km during three days. Only the locations were virtually reversed. Both the advanced *rfo*-Basic version and the *EasyTCP* app worked as expected. The planned short intervals, which cause connection gaps, were encountered, but they were eliminated after three minutes. The complete Sketch to control via the Internet will look like this:

```
// ----------------------------------------------
// ESP8266/ARDUINO : Relay/Relais Pin 7
// H000 - H255      : PWM Pin 6
// DynDNS/GOIP      : Global Control (IoT)
// http://hjberndt.de/soft/ardesp8266goip.html
// ----------------------------------------------

#define LED 7 // Relais/Relay
#define SLD 6 // Helligkeit/Brightness
#define RTIME   180000 //TRIGGER ESP every  RTIME ms (3 Min)
#define RRTIME  900000 //RESTART ESP every RRTIME ms (15 Min)
#define GOTIME  (8*RRTIME) //GOIP-Refresh 2 hour interval
#define URL "goip.de"
#define DOMAIN "GET http://www.goip.de/setip?....\r\n"

char *startup[]={"ATE0","AT+CWMODE=3","AT+CIPMUX=1",
                "AT+CIPSERVER=1","AT+CIPSTATUS"};

unsigned long t0,t1,tg;

void espReset()
{if(analogRead(A0)==0 || millis()>(t1+RRTIME))
 {Serial.println("AT+RST");delay(2000);espReadLine(1000);
  t1=millis();}
 for(int i=0;i<5;i++)
 {Serial.println(startup[i]);espReadLine(0);}
}

void goip()
{char s[80];
 String cmd="AT+CIPCLOSE=4";// disconnect #4 if con
 Serial.println(cmd);espReadLine(0);
 cmd="AT+CIPSTART=4,\"TCP\",\"";cmd+=URL;cmd+="\",80";
 Serial.println(cmd);espReadLine(1000);
 cmd=DOMAIN;sprintf(s,"AT+CIPSEND=4,%d",cmd.length());
 Serial.println(s);  espReadLine(1000);
 Serial.print(cmd);   espReadLine(1000);
```

```
}

void setup()
{pinMode(LED,OUTPUT); Serial.begin(9600);
 t0=millis();t1=t0;tg=t0;
 espReset();
}

char *espReadLine(int ms)
{char t[20],s[80]; //buffer
 if(ms)delay(ms);
 int len=Serial.readBytesUntil('\n',s,sizeof(s)-4); s[len]=0;
 return s;
}

void loop()
{int id,len;char t[10],s[80];
 if(3==sscanf(espReadLine(0),"+IPD,%d,%d:%s",&id,&len,s))
 {switch (s[0])
  {case '0':digitalWrite(LED,LOW);break;
   case '1':digitalWrite(LED,HIGH);break;
   case 'H':analogWrite (SLD,(atoi(strncpy(t,&s[1],3))));break;
  }
  sprintf(s,"AT+CIPSEND=%d,12",id);// +\r\n
  Serial.println(s);espReadLine(0);// message to id
  Serial.println(digitalRead(LED)?"LED is ON ":"LED is OFF");
 }
 if(millis()>(t0+RTIME)){espReset();t0=millis();}
 if(millis()>(tg+GOTIME)){goip();tg=millis();}
}
```

Figure 58: Even an iPod Touch 2nd generation can control via the internet with this structure. In the background a TFT display runs on an Arduino Uno with the connected ESP8266 /1 in AT mode

1.3.8 ESP8266-AT: TIMED CONTROL

IoT may need an exact time stamp - only by this things in the internet can be influenced at the right time, as desired. As global control requires Internet connection, the exact timestamp should be available from there too. So the exact time has to be obtained from the network - due to the system not as easy and the NTP protocol is not very simple too. For the hobbiest or the amateur possibly a reference time as of *http://nist.gov/* provided UTC time of day is sufficient.

Figure 59: Internet time on a TFT display by this method

As the scenario continues to serve a fictitious place in which some things on the internet (IoT) are to be to be controlled. A wireless router, for example, an old Fritzbox, is available there and continuous online. ESP and Fritzbox are familiar and the IP is globally accessible. There is a BASIC structure as it was used in the relay control earlier here above. The previous sketch is again modified accordingly. A present Arduino R3 running at 16 MHz, lags in 6 hours about 30 seconds. Adding 3600000 millis() of 6 seconds, the internal clock runs relatively accurately, without calling an external reference. For about 5 seconds were measured under this approximation in 12 hours. If no computing error, inaccuracy then is 1 hour/year. However, this difference does not seem to be a constant. Probably temperature and type of Sketches impact the internal timer.

To adjust time even manually if necessary (local, no network), an input by smartphone and a TCP client is provided, in addition the Internet time can be manually requested via a TCP client. The time query via the Internet and *ESP8266* requires Internet access. The Arduino is accessible via the *ESP8266* locally by router and the IP 192.162.xxx.xxx and vice versa. The routine time () get the time from the Internet:

```
void time()
{String cmd="AT+CIPCLOSE=4";
 Serial.println(cmd);espReadLine(1000);
 cmd="AT+CIPSTART=4,\"TCP\",\"";
 cmd+="time.nist.gov";cmd+="\",13";
 Serial.println(cmd);
}
```

The time server allows only few calls per interval otherwise usually access is denied. That is why the first attempt for synchronization takes place 30 seconds after start-up, then repeats every 30 seconds until a successful connection is established.

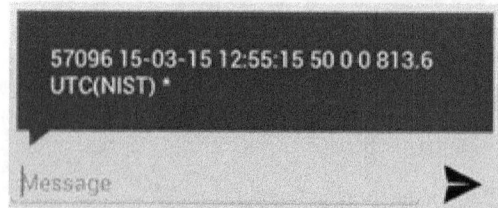

Figure 60: Android-Chat and timestamp

Communication takes place via the ESP8266 connection 4 which is initially closed, and then a second delay is given, to react. Now the new TCP connection 4 on port 13 is established to the time server by URL. In case of a successful request, the server responds with a string of eight parameters that can be extracted. In addition to a 5-digit number there is the time data - as can be seen in the format string. The remaining characters of the line are simply copied into string *t*.

```
...
//Check TimeStamp Daylight
k=sscanf(s,"%5d %2d-%2d-%2d %2d:%2d:%2d %s",
&r5,&y,&m,&d,&hh,&mm,&ss,t);
if(k==8)
{setTime(hh,mm,ss,d,m,y);
 tc=millis();// next hour automatic sync if needed
 tf=0; //sync
}
```

```
//TimeClock UTC to CET
cet=CE.toLocal(now(),&tcr)+2;
...
```

Using *settime()* from the library *Time.h* the 'Unix-clock' of the Arduino is set from 01/01/1970 to the current UTC time, a *tc = 0* inhibits a new automatic time query in 30-second intervals. To get CET from UTC, another library *timezone.h* is required for this conversion. The conversion to Central European Time, taking into account of summer and winter time, is done with *CE.toLocal()*, wherein CE is a time zone and a time cet structure as another variable in addition to the system time in UTC, contains the CET-time. The code is taken from the World Clock Time Zone example of the library, which was reduced to the needs and adjusted accordingly. There the DST rules of other regions can be found.

```
#include <Time.h>
#include <Timezone.h>
//https://github.com/JChristensen/Timezone
//From WorlClock Example
//Central European Time (Frankfurt, Paris)
TimeChangeRule CEST = {"CEST", Last, Sun, Mar, 2, 120};
TimeChangeRule CET = {"CET ", Last, Sun, Oct, 3, 60};
Timezone CE(CEST, CET);
TimeChangeRule *tcr;
time_t cet;
```

The following 130-line-Sketch has no direct output, but only communicates via TCP/IP to a connected device. Dynamic IP and router connection is not included here, although the sketch can be run in conjunction with a 'set' *ESP8266*. This restriction is for reasons of clarity and shortness.

As before the ASCII characters '0' and '1' control a relay or an LED. A 'z' is a time request, a 't' starts an unscheduled time synchronization. The character '8' returns the current operating system and a 'T', the time (UTC) can be set manually in the format of time and date that is returned by 'd'. At the end of the *loop()* a time display could be coded, which then updated approximately every second on the basis of the serial timeout.

```
// -----------------------------------------------
// ESP8266  - Internet-Time
```

```
// Daylight - Timestamp
// Steuern mit WLAN III
// Bezug: http://hjberndt.de/soft/ardesp8266time.html
// ----------------------------------------------
#include <Time.h>
#include <Timezone.h>
//https://github.com/JChristensen/Timezone
//From WorlClock Example
//Central European Time (Frankfurt, Paris)
TimeChangeRule CEST = {"CEST", Last, Sun, Mar, 2, 120};
TimeChangeRule CET = {"CET ", Last, Sun, Oct, 3, 60};
Timezone CE(CEST, CET);
TimeChangeRule *tcr;
time_t cet;

#define LED 7    // Relais/Relay
#define RTIME    180000 //TRIGGER ESP every RTIME ms
#define RRTIME   900000 //RESTART ESP every RRTIME ms
#define CETIME  (3600000)
char *startup[]={"ATE0","AT+CWMODE=3","AT+CIPMUX=1",
                 "AT+CIPSERVER=1","AT+CIPSTATUS"};

volatile int last_id; //sender id for message
unsigned long t0,t1,tc,tf=30000;//first sync after boot

void time()
{String cmd="AT+CIPCLOSE=4";
 Serial.println(cmd);espReadLine(1000);
 cmd="AT+CIPSTART=4,\"TCP\",\"";
 cmd+="time.nist.gov";cmd+="\",13";
 Serial.println(cmd);
}

void setup()
{pinMode(LED,OUTPUT); Serial.begin(9600);
 t0=millis();t1=t0;tc=t0; espReset();
}

void loop()
{int id,len,ix;char t[80],s[80],p[80];
 strcpy(p,espReadLine(0));
 ix=sscanf(p,"+IPD,%d,%d:%s",&id,&len,s);
 if(3==ix)
 {last_id=id;
  switch (s[0])
  {case '0':digitalWrite(LED,LOW);showLed(id);break;
```

```
  case '1':digitalWrite(LED,HIGH);showLed(id);break;
  case '8':espSendLine(id,"show os version");
           Serial.println("AT+GMR");break;
  case 'T'://Settime "T00:00:00D01.01.2000" UTC
           {int y,m,d,hh,mm,ss;
            ix=sscanf(s,"T%2d:%2d:%2dD%2d.%2d.%4d",
                      &hh,&mm,&ss,&d,&m,&y);
            if(ix==6)
            {setTime(hh,mm,ss,d,m,y);
             espSendLine(id,"UTC Time Set.");
            }
           }
           break;
  case 't':espSendLine(id,"try sync.");time();break;
  case 'z':sprintf(t,"%02d:%02d:%02d",
           hour(cet),minute(cet),second(cet));
           espSendLine(id,t);
           break;
  case 'd':sprintf(t,"T%02d:%02d:%02dD%02d.%02d.%04d",
           hour(cet),minute(cet),second(cet),
           day(cet),month(cet),year(cet));
           espSendLine(id,t);
           break;
  default:showLed(id);
 }
}
//Check prompt
len=strlen(p);
if(p[0]=='>' && len>5)
{for(int i;i<len;i++)if(p[i]<32)p[i]=0;
 espSendLine(last_id,p);//reply to last sender
}
if(millis()>(t0+RTIME)){espReset();t0=millis();}
if(millis()>(tc+CETIME))
{if(timeStatus()<2)time();//ONE SYNC
 else adjustTime(6);        // sec per hour
 tc=millis();               //next hour
}
if(tf>0)if(millis()>(t0+tf))
{tf+=tf; time();}//first sync after tf
// CLOCK
// drawClockHands(cet); //not included
}

void showLed(int id)
{ char s[80];
```

```
    sprintf(s,"AT+CIPSEND=%d,12",id);// Message to id
    Serial.println(s);espReadLine(0);
    Serial.println(digitalRead(LED)?"LED is ON ":"LED is
OFF");
}

void espSendLine(int id, char *t)
{char s[80];
 sprintf(s,"AT+CIPSEND=%d,%d",id,strlen(t)+2);
 Serial.println(s); Serial.println(t);// Message to id
}

char *espReadLine(int ms)
{int k,r5,y,m,d,hh,mm,ss;char t[80],s[80]; //buffer
 if(ms)delay(ms);
 int len=Serial.readBytesUntil('\n',s,sizeof(s)-4);
 s[len]=0;
 //Check TimeStamp Daylight
 k=sscanf(s,"%5d %2d-%2d-%2d %2d:%2d:%2d %s",
          &r5,&y,&m,&d,&hh,&mm,&ss,t);
 if(k==8)
 {setTime(hh,mm,ss,d,m,y);
  tc=millis();// automatic sync if needed
  tf=0;
 }
 //TimeClock UTC to CET
 cet=CE.toLocal(now(),&tcr)+2;
 return s;
}

void espReset()
{if(analogRead(A0)==0 || millis()>(t1+RRTIME))
 {Serial.println("AT+RST");delay(2000);
 espReadLine(1000); t1=millis();}
 for(int i=0;i<5;i++)
 {Serial.println(startup[i]);espReadLine(0);}
}
```

In this short solution the result is returned to the last ID only, not to all connected control units.

1.4 DIGISPARK

This board sometimes is called "The smallest Arduino". An ATtiny85 microcontroller and a cleverly constructed board, designed as a USB plug allows to solve many tasks or problems 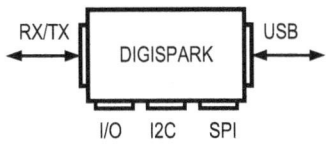 without having to get the "big" small Arduino from the drawer.

At *digistump.com* the details and specifications of this Kick-starter project are mentioned:

- Support of the Arduino IDE 1.0+ (OSX /Win/Linux)
- Power supply via USB or external source: 5 V or 7-35 V
- On board Voltage regulator
- Built-in USB
- 6 I/O pins (2 used by USB, if the sketch communicates via USB, otherwise all 6 ports can be used for own tasks)
- 8 kB Flash memory (about 6 kB including bootloader)
- I²C and SPI (vis USI)
- PWM to 3 pins (more possible with software PWM)
- ADC to 4 pins
- Power LED and test/status LED

Figure 61: USB host cable (OTG) and Digispark

1.4.1 DIGISPARK: IDE

As the Arduino-IDE 1.6 and later integrates other boards in a simple way the Digispark can be added by entering an URL in the settings to download the necessary files from the File menu and then set the board in Tools/Board. For a Digispark that URL at the time of writing down this text was:

https://raw.githubusercontent.com/digistump/arduino-boardsindex/master/package_digistump_index.json

There may be reasons to install an own (Arduino) IDE for the Digispark. Digistump supplied or supplies the file:

https://github.com/digistump/DigistumpArduino/releases/downloa d/v1.5.8c/DigistumpArduinoInstall1.5.8C.exe

This setup includes all the necessary drivers for the Arduino too. Further details on this subject can be found or were found at:

http://www.cboden.de/mikro-controller/digispark/20mikrocontroller/43-erste-schritt-mit-dem-digispark

In the German language or in English at:

https://digistump.com/wiki/digispark/tutorials/connecting

Having done the necessary preparations, a first test can be performed. At "Tools" the "Board" Digispark should set to be running at 16.5 MHz / internally. All other settings are not of interest.

1.4.2 DIGISPARK: BLINK

As using an Arduino, the LED on the Digispark is to be lit by the first sketch. This uses the example Blink from the File menu */examples/basics*. This classic Arduino sketch still codes pin 13 as output as fixed number in the sketch. However, the ONBOARD LED of the Digispark is connected to pin 1. Replacing the 13 to 1 in all of the three places, syntax can be checked with the check icon below the File menu.

```
/*
Blink
Turns on an LED on for one second, then off for one se-
cond, repeatedly.
Most Arduinos have an on-board LED you can control. On
the Uno and Leonardo, it is attached to digital pin 13.
If you're unsure what pin the on-board LED is connected
to on your Arduino model, checkthe documentation at
http://arduino.cc

This example code is in the public domain.

modified 8 May 2014
by Scott Fitzgerald
*/


// the setup function runs once when you press reset or
power the board
void setup() {
  // initialize digital pin 13 as an output.
  pinMode(1, OUTPUT);
}

// the loop function runs over and over again forever
void loop() {
  digitalWrite(1, HIGH);
// turn the LED on (HIGH is the voltage level)
  delay(1000);
// wait for a second
  digitalWrite(1, LOW);
// turn the LED off by making the voltage LOW
  delay(1000);
// wait for a second
}
```

If the following or similar output is shown in the lower IDE window, there were no problems.

```
Der Sketch verwendet 650 Bytes (10%) des Pro-
grammspeicherplatzes. Das Maximum sind 6.012 Bytes.

Globale Variablen verwenden 9 Bytes des dynamischen
Speichers.
```

Dramatically can be seen; there is only about 6K of memory space. The upload of a sketch to the Digispark is started by Ctrl+R. At this stage the Digispark should not be connected to the computer. If the IDE is ready for uploading, the following additional text will appear in the message window:

```
Running Digispark Uploader...

Plug in device now... (will timeout in 60 seconds)
```

Now the Digispark has to be connected via USB. Windows plays the sound of an added USB device; next the fast upload is done. In conclusion, the driver closes down. Upload is now complete and the LED should flash every second!

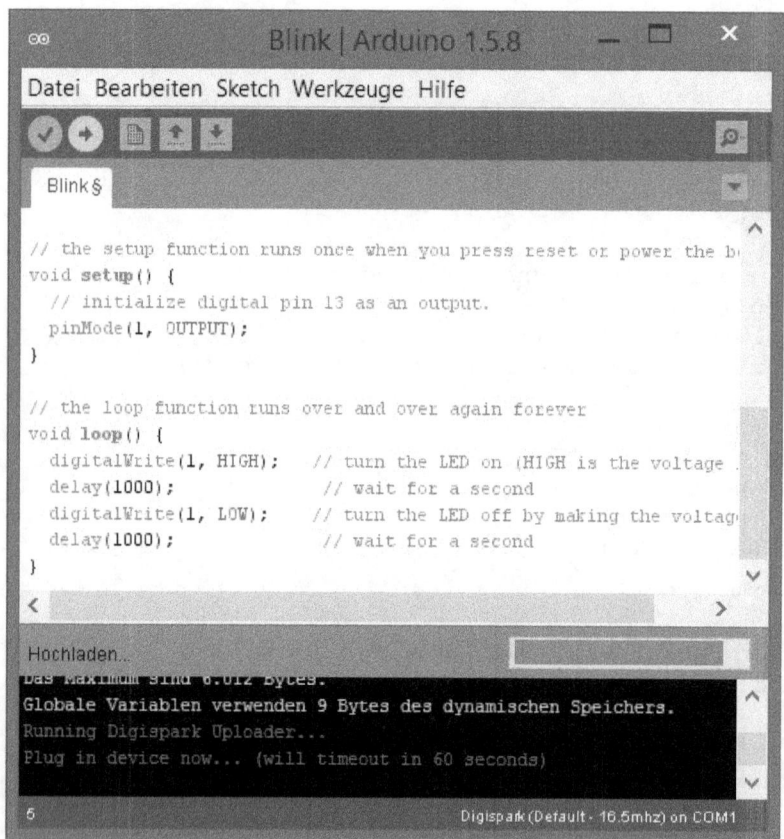

Figure 62 Digispark runs Arduino sketch

1.4.3 DIGISPARK: USB-KEYBOARD

The special feature of the Digispark is the board shape with USB connector and the ability to behave as a HID USB device. Human Interface Devices need no special drivers on the major operating systems and can thus widely be used immediately. A Digispark can be programmed as a joystick, mouse or keyboard. The next example enters an analogue voltage measurement directly as a keyboard into any application on any OS. Under *Examples/Digispark/DigisparkKeyboard* there is the sketch "Keyboard.ino". If all the comments are removed, three main parts remain.

```
void loop()
{
  DigiKeyboard.sendKeyStroke(0);
  DigiKeyboard.println("Hello Digispark!");
  DigiKeyboard.delay(5000);
}
```

In this endless loop first the Digispark is shaken awake, and then it writes the adapted "Hello World" as an connected USB keyboard followed by a 5 second wait.

For testing this sketch, it has to be uploaded, as earlier the blink example. During upload/compilation the writing cursor should be set behind the last brace in the editor. After transfer, the driver logs off as usual, the Digispark remains unchanged connected to the USB port, the PC will recognize the Digispark after a while as a USB-keyboard. Then the sketch starts and if the cursor is still behind the last brace there will appear "Hello Digispark" in a 5 second interval. All this can be tested on various other USB host-capable devices.

1.4.4 DIGISPARK: MEASURING VOLAGES ADC

Analogue inputs deviate from those of the digital connections as
https://digistump.com/wiki/digispark/quickref
says:

Digital 2 is analog (ADC channel) 1
Digital 3 is analog (ADC channel) 3
Digital 4 is analog (ADC channel) 2
Digital 5 is analog (ADC channel) 0

Due to its compact appearance, the double function of the connections
has to be considered, as a result, the connections behave differently depending on the function.

All pins can be used as Digital I/O
Pin 0 → I2C SDA, PWM (LED on Model B)
Pin 1 → PWM (LED on Model A)
Pin 2 → I2C SCK, Analog In
Pin 3 → Analog In (also used for USB+ when USB is in use)
Pin 4 → PWM, Analog (also used for USB- when USB is in use)
Pin 5 → Analog In

Under these conditions, an USB measuring device or the measured value
input via a USB keyboard cannot use pin 3 (ADC 3). Pin 2 is available.
However, if an OLED display is to be connected via I2C, as described below, this connector is used as a serial clock (SCK). With very few changes,
the analogue value of pin 2 in the range 0 to 1023 is output in the following sketch. The voltage at the ADC is digitized with a 10-bit resolution.

```
#include "DigiKeyboard.h"
#define LED 1
#define ADC 2

void setup()
{pinMode(ADC, INPUT);
}

void loop()
{DigiKeyboard.sendKeyStroke(0);
 DigiKeyboard.println((int)analogRead(1));
 DigiKeyboard.delay(5000);
}
```

Using Arduino-IDE:

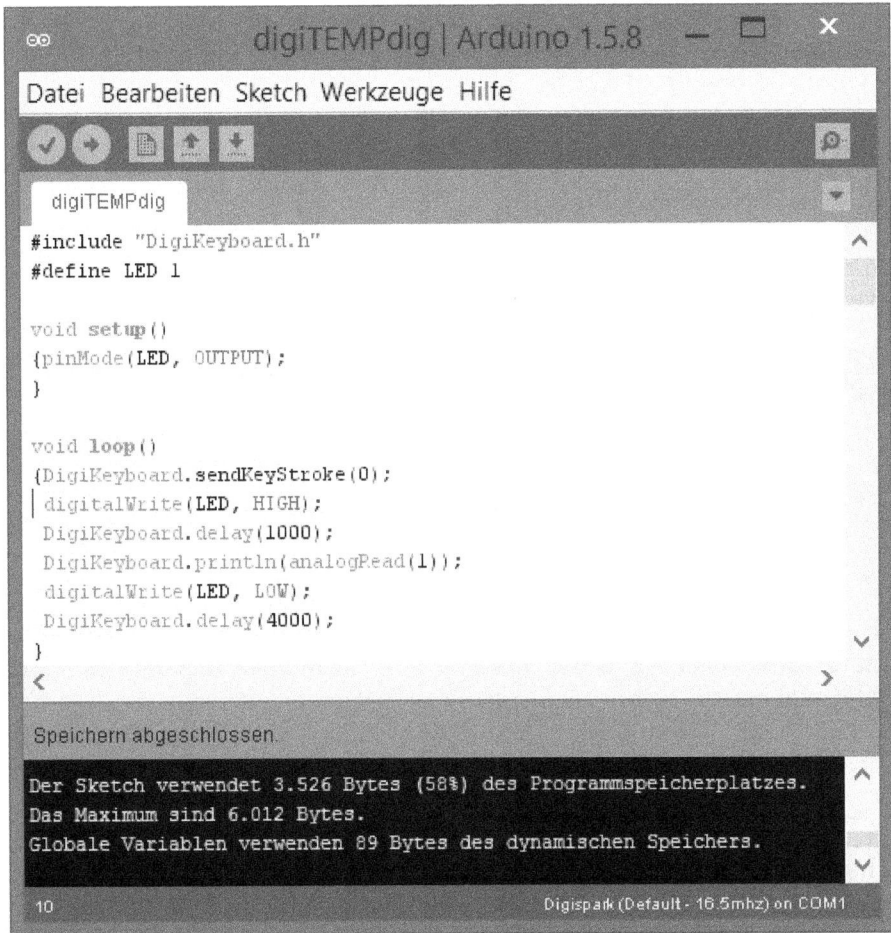

Figure 63: Digispark as keyboard providing analogue voltage values

The switching of the LED is optional. The conversion into a voltage value could be done in a spread sheet or any application of that kind.

1.4.5 DIGISPARK: VALUES AND POINT AND COMMA

If no external conversion option is available, this can be done by the sketch. Depending on the target application decimal point or comma (German) are in demand. The conversion for a 5 Volt voltage range at pin 2 would be x = adc / 1023 * 5.

```
x = adc/1023 times 5
```

This could accomplish the following line:

```
DigiKeyboard.println(analogRead(1)/1023*5);
```

To renounce floating point arithmetic in this lising in terms of the low memory space integer operations are more effective. The middle line is replaced by

```
long l=analogRead(1)*5;
DigiKeyboard.print(l/1023);
DigiKeyboard.print(',');
DigiKeyboard.println(l%1023);
```

Using division and modulo the voltage decimal places are calculated. The characters in between is the comma, but this can be changed if needed to a decimal point.

1.4.6 DIGISPARK: TIME AND VOLTAGE MEASUREMENT

The elapsed time can be tranferred also to get a voltage-time chart. The function *millis()* returns the number of milliseconds elapsed since start of the sketch. The simplest implementation may be:

```
long l=analogRead(1)*5;
DigiKeyboard.print(millis()/1000);
DigiKeyboard.print("\t");
DigiKeyboard.print(l/1023);
DigiKeyboard.print(',');
DigiKeyboard.println(l%1023);
```

The measurement of the voltage is followed by a time output in seconds. The sequence "\t" outputs a tab character. After conversion is done, as

shown above, the corresponding measuring point can be "typed" - and
being represented graphically as required in any application and any OS

Figure 64: Decimal voltage measurement

1.4.7 DIGISPARK: TEMPERATURE USING LM35

An LM35-sensor allows simple temperature measurement. Its three connections are power supply V_{cc}, ground Gnd and the measured value output. The simplest use is if only positive temperatures in °C are to be measured. Then a simple connection of the three pins is sufficient.

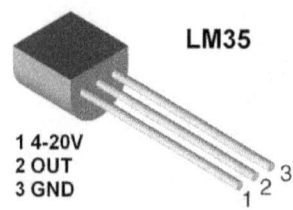

Figure 65: Analoger Temperatursensor LM35

Since the sensor produces only a small current load, the power supply by means of digital output is to take place on Digispark. The ground connection is to be made switchable in this example. Corresponding lines or connections are defined in the listing:

```
#include "DigiKeyboard.h"
#define LED 1
#define VCC 1
#define ADC 2
#define GND 0
```

Pin 0 is ground supply, pin 1 is power supply with simultaneous display of the state by the built-in LED. The measurement is performed again on pin 2. This arrangement of the LM35 with little bending of the legs can be connected directly to the Digispark. In *setup()* defines for the inputs and outputs, as well as placing the power supply for the sensors is done. In addition, the switch over to the internal reference of 1.1 Volts is done here. Thus, the digital value 1024 corresponds to a voltage of 1.1 Volts. Taking into account the inaccuracy of the reference and the fact that there has to take place no calibration here; the accuracy for many applications could be ok. A digital value of e.g. 256 then is equal to 25.6 °C.

```
void setup()
{pinMode(VCC, OUTPUT);
 pinMode(ADC, INPUT);
 pinMode(GND, OUTPUT);
 digitalWrite(VCC, HIGH);
 digitalWrite(GND, LOW);
```

```
analogReference(INTERNAL1V1);
}
```

There is still a 5 seconds-interval to measure. In the loop now the dependent sensor converting is done of the analogue voltage, since the temperature in °C corresponds to the voltage in mV.

```
void loop()
{DigiKeyboard.sendKeyStroke(0);
 digitalWrite(VCC, HIGH);
 DigiKeyboard.delay(1000);
 unsigned long l=analogRead(1);
 DigiKeyboard.print(l/10);
 DigiKeyboard.print(',');
 DigiKeyboard.println(l%10);
 digitalWrite(VCC, LOW);
 DigiKeyboard.delay(4000);
}
```

1.4.8 DIGISPARK: ON BOARD TEMPERATURE

The electronics of the Digispark and the Attiny85 contain an internal temperature sensor, which is very suitable for smaller experiments. Because the electronics heat up for operational reasons, minor corrections may be necessary. How to address the sensor is found at:

https://digistump.com/wiki/digispark/quickref

The routine *temp()* returns the temperature in °C.

```
int temp()
{analogReference(INTERNAL1V1);
 int raw = analogRead(A0+15);
 raw -= 7; // used to calibrate
 int in_c = raw - 273; // celcius
 analogReference(DEFAULT);
 return in_c;
}
```

The reference is switched, then a temperature value is read in degrees Kelvin of analogue line A0+15, corresponding to the calibrated chip -7 K and returned as an integer value in Celsius. The measurement loop then looks like this:

```
void loop()
{DigiKeyboard.sendKeyStroke(0);
 digitalWrite(LED, HIGH);
 DigiKeyboard.println(temp());
 DigiKeyboard.delay(1000);;
 digitalWrite(LED, LOW);
 DigiKeyboard.delay(4000);
 }
```

The Digispark LED signals the measurement. After starting the sketch, the on-board temperature is transmitted via the USB port to a smartphone or a tablet as keystrokes. There is still enough memory available for transferring the measurement time stamp.

1.4.9 DIGISPARK: HEATING-/LOADING CURVE

Temperature ranges or control circuits of different orders can be electrically simulated by appropriate capacitances and resistances. The heat capacity finds its electrical analogy in the form of a capacitor and the heat resistance as a transitional value is by an electrical resistance the corresponding analogy. The voltage profile of the charging curve of an electrical capacitor corresponds to the heating curve of a temperature range of 1st order and is described by the following equation:

$$u(t) = u_0\left(1 - e^{-t/\tau}\right)$$

In practice there are more or less differences by various measurement errors from this theoretical ideal curve. Next is to try to measure such a curve with this "smallest Arduino in the world" and to transfer these data to an application keying in measurement values. A presentation and analysis should then, if needed, for reasons of space be done on the receiving smartphone or tablet.

As indicated above using the temperature sensor LM35, a digital output that is switched from 0 volts to 5 Volts serves as a controllable power source. The analogue input of pin 2 (ADC 0) refers to the voltage of the capacitor. The listing below contains the concrete connections.

If such a function using $\tau = 1$ for 5 seconds is graphically illustrated, the result is a graph like this:

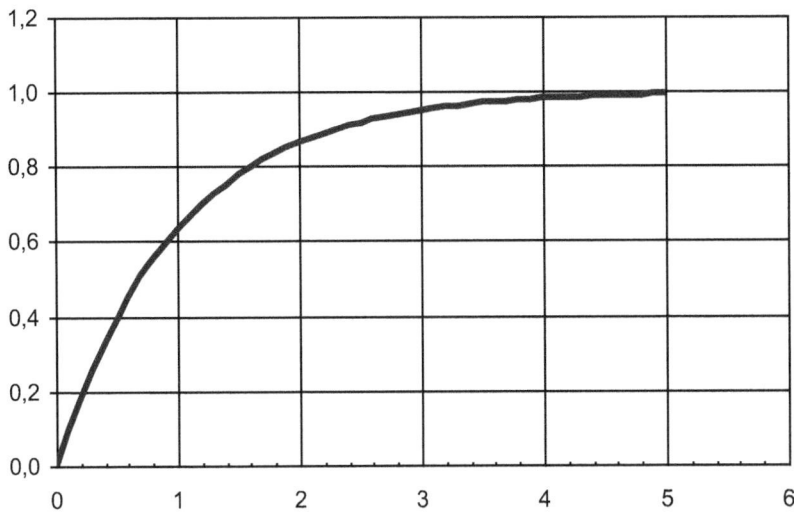

Figure 66: Heating-/Loading Curve

To transfer measurement data as a voltage value with decimal places in USBKeyboard mode, the routine *Key()* is used to make the sketch look more simple. Furthermore, no floating-point arithmetic is used. The overall process is as follows:

- Output 5 V
- repeat
  - o voltage measurement
  - o transfer measurement
  - o delay
- until the capacitor is nearly full
- Output 0 V
- repeat
  - o voltage measurement
  - o delay
- until the capacitor is almost empty
- wait and discharge for 5 seconds

The implementation for a Digispark is designed in the following syntax:

```
#include "DigiKeyboard.h"
```

```
#define VCC 1
#define ADC 2
#define GND 0

void setup()
{pinMode(VCC, OUTPUT);
 pinMode(ADC, INPUT);
 pinMode(GND, OUTPUT);
 digitalWrite(GND, LOW);
}

void Key(long value)
{DigiKeyboard.print(value/1023);
 DigiKeyboard.print(',');
 DigiKeyboard.println(value%1023);
}

void loop()
{long wert;
 DigiKeyboard.sendKeyStroke(0);
 digitalWrite(VCC, HIGH);
 do
  {wert=analogRead(1)*5;
   Key(wert);
   DigiKeyboard.delay(500);
  }while (wert/5>50);
 DigiKeyboard.sendKeyStroke(0);
 digitalWrite(VCC, LOW);
 do
  {wert=analogRead(1)*5;
   DigiKeyboard.delay(200);
  }while (wert/5<985);
 DigiKeyboard.delay(5000);
}
```

During the loading phase, while output is set to 5 volts, the connected LED is lit. Without measured value the program behaves as a flashing sketch *blink*, in which the frequency is determined by multiplying C times R (tau). With the components C = 1000 microfarads and R = 3 kΩ arise corresponding curves.

1.4.10 DIGISPARK: DISCHARGE-/COOLING CURVE

Switching off a heat source, the temperature profile often corresponds to an exponential function. It looks the same as the voltage source of the *R/C*-circuit is switched off. To get the decay curve only, the *key* statement needs to be moved in the lower loop. The result at 1000 microfarads and 2000 Ω and a measurement interval of roughly 0.5 seconds has the following plot:

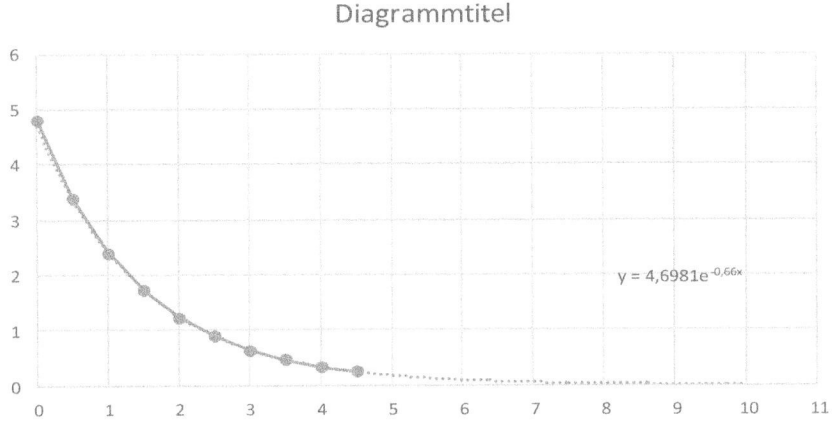

Figure 67: Cooling curve or discharging

1.4.11 DIGISPARK: SOFTSERIAL

Using a Digispark, data can also be transferred serially. Using an Arduino *Serial.print* is a popular tool to make results of a program sequence visible. Using a Digispark this is not possible because the lack of a USB to serial module. Nevertheless, the digital terminals can be connected as the *Rx/Tx* lines at TTL-level. Thus, the Digispark can connect in this way to corresponding components. Some modules with TTL serial are:

- Arduino
- Esp8266
- FTDI-adapter
- GPS-module
- HC-Bluetooth-adapter

1.4.12 DIGISPARK: BLUETOOTH

In [2] is shown how a LED can be controlled by Bluetooth using the Arduino. This sketch for the Arduino is to be modified with minimal adjustments for a Digispark, producing the same result. On the smartphone nothing has to be changed. The Bolutek Bluetooth adapter, there used, is replaced by the widespread *HC06* module.

The first connection is to be tested with a "Hello Digispark". If this message can be received wireless on a smartphone via Bluetooth, the adaptation of the old Arduino Sketch is next. With RX at pin 2 and TX at pin 3 the USB functionality is not possible anymore, but needed any longer. A particularly for the Digispark available "TinyPinChange" library is included as well as the Arduino *SoftSerial* library. After establishing the connections, a soft Serial object is created.

```
#include <SoftSerial.h>
#include <TinyPinChange.h>
#define  RX     2
#define  TX     3

SoftSerial mySerial(RX, TX);
#define Serial mySerial

void setup()
{Serial.begin(9600);
}

void loop()
{Serial.println("Hallo Digispark");
 delay(1000);
}
```

The *HC06* is set to 9600 baud by default and that is why the serial interface is set to this speed. The rest is repeating something every second. As the Digispark is powered via its USB connector, the *HC06* can obtain its power from there. So there are four wires connecting Digispark and *HC06*.

Digispark	HC06
5V	Vcc
Gnd	Gnd
Pin 2 (RX)	TX
Pin 3 (TX)	RX

1.4.13 DIGISPARK: CONTROL USING BLUETOOTH ON/OFF

In the year 2014, the experiment was done using the Arduino and a Bolutek Bluetooth adapter. This section

http://www.hjberndt.de/soft/android/indexbolu.html

was added to [2]. Now the hardware is again drastically reduced in price. The Sketch for Arduino can now be reused in essence. The original source of the Arduino experiment was or is at:

http://english.cxem.net/arduino/arduino5.php

Only the LED is addressed through pin 1 using Digispark.

```
#include <SoftSerial.h>
#include <TinyPinChange.h>
#define   RX    2
#define   TX    3

SoftSerial mySerial(RX, TX);
#define Serial mySerial

char incomingByte; // incoming data
int LED = 1; // LED pin

void setup()
{Serial.begin(9600); // initialization
 pinMode(LED, OUTPUT);
 Serial.println("Press 1 to LED ON or 0 to LED OFF...");
}

void loop()
{if (Serial.available() > 0)
 {// if the data came
  incomingByte = Serial.read(); // read byte
  if(incomingByte == '0')
  {digitalWrite(LED, LOW); // if 1, switch LED Off
    Serial.println("LED OFF. Press 1 to LED ON!"); //
print message
  }
  if(incomingByte == '1')
  {digitalWrite(LED, HIGH); // if 0, switch LED on
   Serial.println("LED ON. Press 0 to LED OFF!");
  }
```

```
  }
}
```

The original source code is modified by deleting one character from the 13, so pin 13 becomes pin 1. The six first lines provide serial communication and Digispark compatibility.

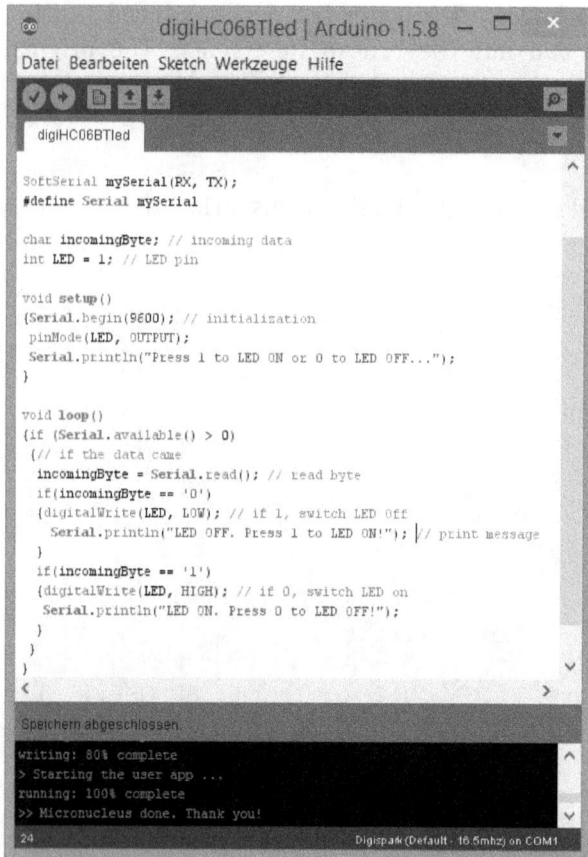

Figure 68:
LED-Control using
Bluetooth and
Digispark

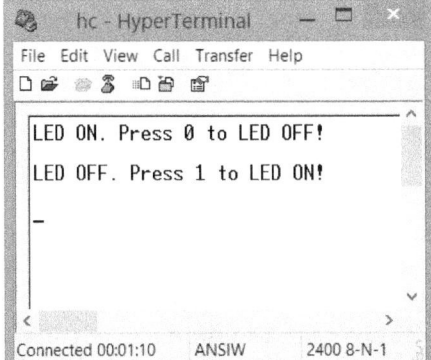

Figure 69: Hyperterminal and Digispark-Bluetooth

The old HyperTerminal works fine using Bluetooth. The transmission rate is automatically detected; it has not been set here. Only the number of the COM port must be known. This is obtained in the Bluetooth settings using Windows. Windows 8.1 is not that fast connecting to the HC06. The first connection attempt might not be successfully, a repeated try will help. *RealTerm* gets a connection mostly on the first try, albeit after a certain memorial break.

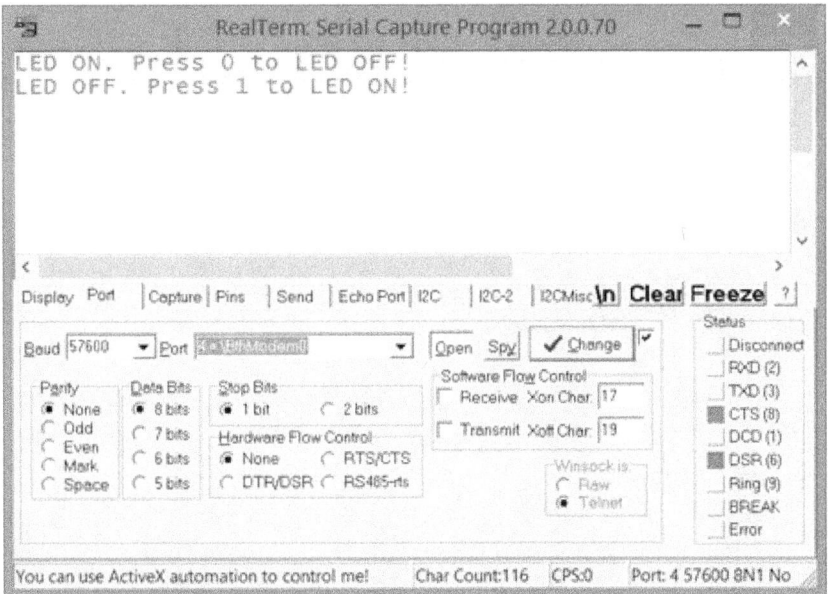

Figure 70: Digispark and Bluetooth in RealTerm

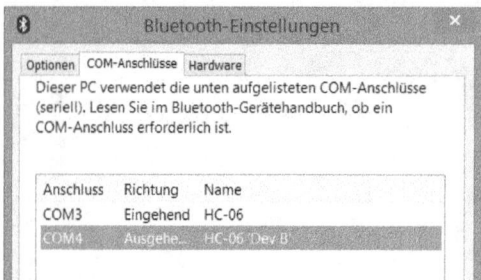

Figure 71: Finding Bluetooth-COM in Win 8.1

RealTerm Windows copes with Bluetooth, but can do much more. A very brief outline can be found in Section 2 "Software Elements" in this book. There some explanations concerning *VBScript* can be found too.

For the Android smartphone or tablet, there are many free applications that allow the control using Bluetooth this way. Some of them have been successfully used in [2].

```
' Writing Data to a Text File

Const ForAppending = 8

Set objFSO = CreateObject("Scripting.FileSystemObject")
Set f = objFSO.OpenTextFile("COM4:", 8, True)
While True
      WScript.Sleep 2000
      f.WriteLine("1"):WScript.Echo Time
      WScript.Sleep 2000
      f.WriteLine("0"):WScript.Echo Time
Wend
f.Close
```

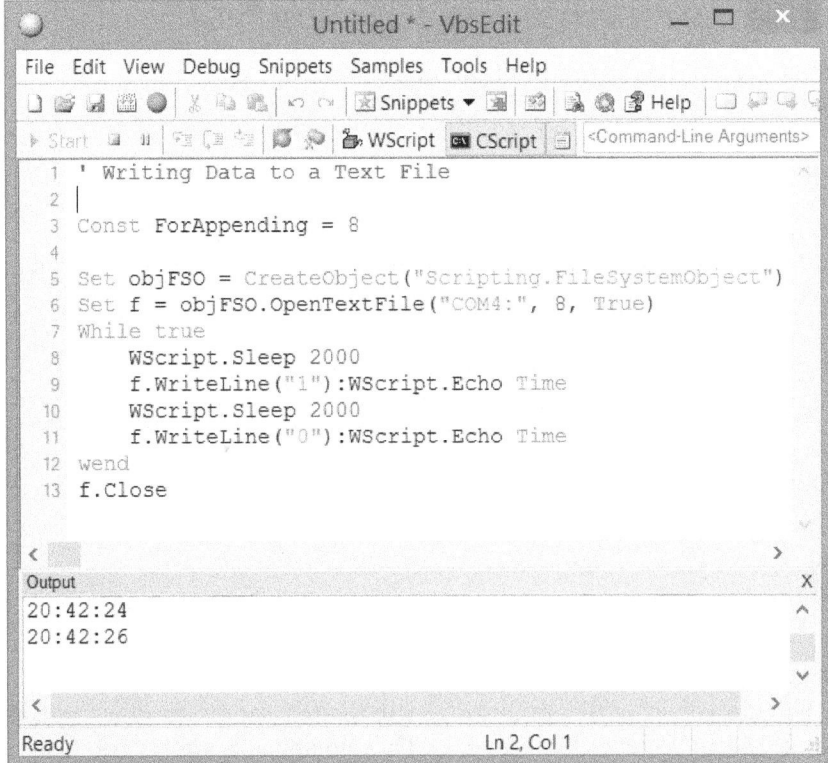

Figure 72: Blink-Control in VBScript

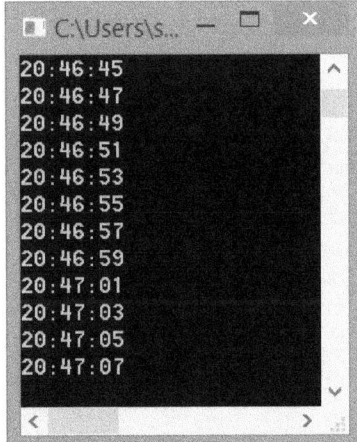

Figure 73: VBS and console-output

1.4.14 DIGISPARK: CONTROL USING BLUETOOTH PWM

An excerpt from the documentation on Digistump says:

All pins can be used as Digital I/O
Pin 1 → PWM (LED on Model A)

So the LED can be controlled in brightness continuously. A first test is to let the brightness rise slowly. The sketch can remain short.

```
void setup() {} // nichts zu tun

void loop()
{for (int i=0;i<256;i++)analogWrite(1,i);
}
```

```
Der Sketch verwendet 606 Bytes (10%) des Pro-
grammspeicherplatzes. Das Maximum sind 6.012 Bytes.

Globale Variablen verwenden 9 Bytes des dynamischen Speichers.
```

However, the result is disappointing. The sketch seems to work not as expected. Taking a closer look reveals: The Digispark is simply too fast!

Inserting a delay of 10 ms everything is working as expected.

```
void loop()
{for (int
i=0;i<256;i++){analogWrite(1,i);delay(10);}
}
```

Or a little more smoothly using these lines:

```
void loop()
{for (int i=0; i<256; i++)
 {analogWrite(1,i);delay(10);}
 for (int i=254;i>0;i--)
 {analogWrite(1,i);delay(10);}
}
```

A control by phone or tablet requires some connection between the two devices. The simplest wireless connection provided to the Digispark is *SoftSerial* and a *HC06* Bluetooth adapter. On the side of Digispark this requires the integration of the serial routines for receiving the control commands. On a smartphone/tablet a Bluetooth terminal can be the endpoint app - as elsewhere. Since only two characters no longer are sufficient for this control, another receiver technology is needed. The brightness has to be controlled by three digits, which form values from 000 to 255. The reason for this choice is the free Android app BlueTooth Serial Commander from the PlayStore that is to be used on an Android smartphone.

The section *control with Digispark* presents a method for entering values terminated with a line feed. However, the Android app is not sending a newline character when using the slider.

1.4.15 DIGISPARK: CDC: SERIALUSB

Using the Digispark it is possible to realize a virtual serial interface via USB. At

https://digistump.com/wiki/digispark/tutorials/digicdc

the corresponding versions can be found: "The DigiCDC library allows the Digispark or Digispark Pro to appear to a computer as a Virtual Serial Port when connected by USB. This makes it appear just like a standard Arduino and allows the use of the Serial monitor built into the Arduino IDE."

Thus, a Digispark should behave like a kind of *FTDI-adapter* and can, like an Arduino, communicate serially via a USB interface. This works well for own tests on Windows 7/32, but not for Winows 8.1/32. The smartphone with Android 4.x has no issues; this is why a corresponding test program is listed here.

Under *File/examples/DigisparkCDC/* one finds the example CDC_LED which is accepted here without comments.

```
#include <DigiCDC.h>
#define LED 1
void setup()
```

```
{SerialUSB.begin();
 pinMode(LED,OUTPUT);
}

void loop()
{if(SerialUSB.available())
 {char input = SerialUSB.read();
  if(input == '0') digitalWrite(LED,LOW);
  else if(input == '1') digitalWrite(LED,HIGH);
 }
 SerialUSB.delay(100);
}
```

This sketch controls the Digispark LED, and thus is a third variant of the small board. The serial interface is called SerialUSB and should behave like the original interface. A transmission rate is not required, so the transfer parameter is absent in *SerialUSB.begin()*. It is only checked in the example, if a "1" or a "0" is received to turn on or turn off the LED correspondingly.

If the Android USB Serial Monitor Lite is installed from the Play Store on the device, the test can be done. A host adapter connects the Digispark to the USB connector of a smartphone.

Figure 74: If is agreed, the LED can be switched by inputting a 0 or 1.

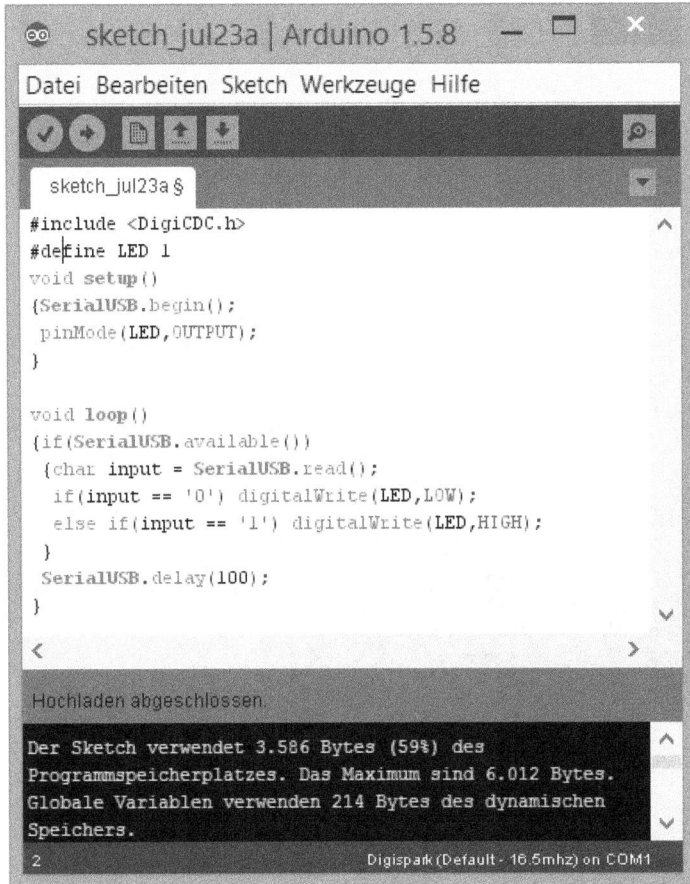

Figure 75: Virtual Digispark USB-interface in the Arduino IDE

The limits of this method for a given hardware can be observed in the output every second. The two-digit seconds arrive slowly. By an additional line feed the output is refused.

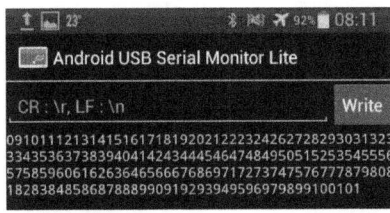

Figure 76:
USB/Smartphone receiving Digispark
data every second

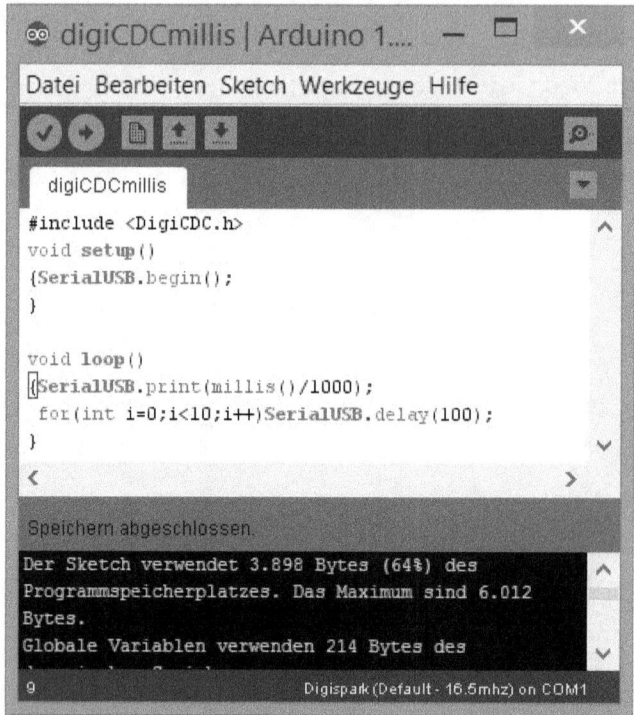

Figure 77: Timed output

The smartphone connected after 9 seconds, due to the necessary manual operation of the app. Thereafter, data arrives every second.

Due to the technical and practical issues using the Digispark1 communication this way, it is not pursued.

1.4.16 DIGISPARK: BLUETOOTH-KEYBOARD

Looking at the possibilities of the Digispark might result in the idea of combining the USB keyboard options with those of the SoftSerial possibilities. At least two people had this idea and wanted Bluetooth (HC module) and SoftSerial to operate the Digispark almost as a Bluetooth keyboard with USB connection. However, problems related to interrupt connections in both processes ceased. The original libraries allowed only either USB keyboard or SoftSerial. At the Forum (JRios «Reply # 7 on: February 05, 2016, 02:02:52 pm»)

https://digistump.com/board/index.php?topic=1956.0 they met and created a new library for exactly this problem. This work is called SoftSerialINT0 and is available at on github:

https://github.com/J-Rios/Digispark_SoftSerial-INT0

If this library is manually added to libraries (for example *C: \ Users\... \ Documents\ Arduino\ libraries*), one of the examples can be tested. Here's a simple Bluetooth echo:

```
#include <SoftSerial_INT0.h>
#define P_RX 2
#define P_TX 1
SoftSerial Bluetooth(P_RX, P_TX);

void setup()
{Bluetooth.begin(9600);
}

void loop()
{if(Bluetooth.available())
 Bluetooth.write(Bluetooth.read());
 delay(100);
}
```

The *TinyPinChange* library is obsolete. This simple test only checks the functionality of the modified *SoftSerial* library. As terminal for RX pin 2 is used, pin 1 is connected as TX. By connecting Digispark to the HC-06 module crosswise, thus TX with RX and vice versa, all characters should appear twice in a terminal app.

Figure 78: Own echo in BlueTerm (Android) on a phone

Now the connections no longer interfere and the second example with the name *Bluetooth keyboard* can be loaded. The second library to include is the unchanged *DigiKeyboard* for use as USB keyboard. After uploading, the keyboard detection and the Bluetooth connection, now there is a device allowing sending keystrokes as USB keyboard (e.g., a Windows-Tablet) from another device (e.g. a smartphone) via Bluetooth for control purposes. In the example, the three capital letters A, B and C are intercepted and replaced by corresponding sequences.

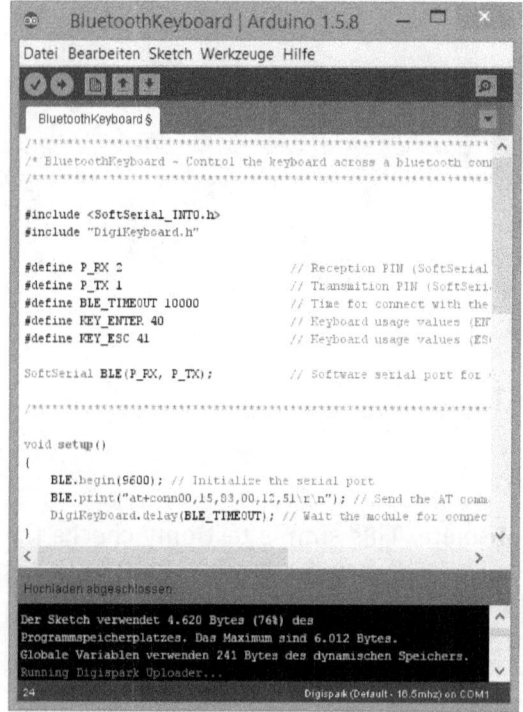

Figure 79:
Digispark-
BluetootKeyboard in IDE

1.4.17 DIGISPARK: I²C-OLED-DISPLAY

A tiny OLED display with I²C control fits qua geometry extremely well to a tiny Digispark1. The control using only two lines invites to bring simple output directly without phone or tablet to be displayed. As a first test, the voltage at pin 2, and the elapsed time will be displayed. Parts of code are identical to the output via USB keyboard. The OLED implementations come from the example Sketch *DigisparkOLED*.

Figure 80: OLED Display and Digispark (Radio controlled Clock project)

The libraries to be included are called

```
#include <DigisparkOLED.h>
#include <Wire.h>
```

Now decisions concerning pin-use (connections) have to be taken. For the I²C lines pin 0 is for SDA (data) and pin 2 for SCL (clock) set by the library. The voltage is to be measured at pin 3 (ADC3). Thus the sketch looks like this:

```
void setup()
{oled.begin();
 oled.setFont(FONT8X16);
 oled.clear(); //all black
 pinMode(3, INPUT);
}

void loop()
{delay(500);
 oled.setCursor(0,0);
 oled.print(millis()/1000);
 delay(500);
 oled.setCursor(0, 4);
 long l=analogRead(3)*5; //ADC3
 oled.print(l/1023);
 oled.print(',');
 oled.print(l%1023);
 oled.print(" V");
}
```

The connection of the 1306OLED is done by only four wires:

Digispark	Oled1306 I^{2}C
Gnd	Gnd
5 Volts	Vcc
Pin 0	SDA
Pin 2	SCL
Pin 3	Input

The first line increments every second, the second line is left blank, in the third line, the voltage at pin 3 is measured in volts. As the range is fixed to 5 Volts, so a resolution of 10 bits correspondent to about 5 mV. As some space in Digispark memory is left, in another variant, the internal temperature could be displayed in °C. All recently testing is a little far from the title of this book, because the smartphone is left out. Combining the code at Digispark as a USB keyboard it should be possible, if required, although there were issues using USB Keyboard an internal temperature measurement in first tests.

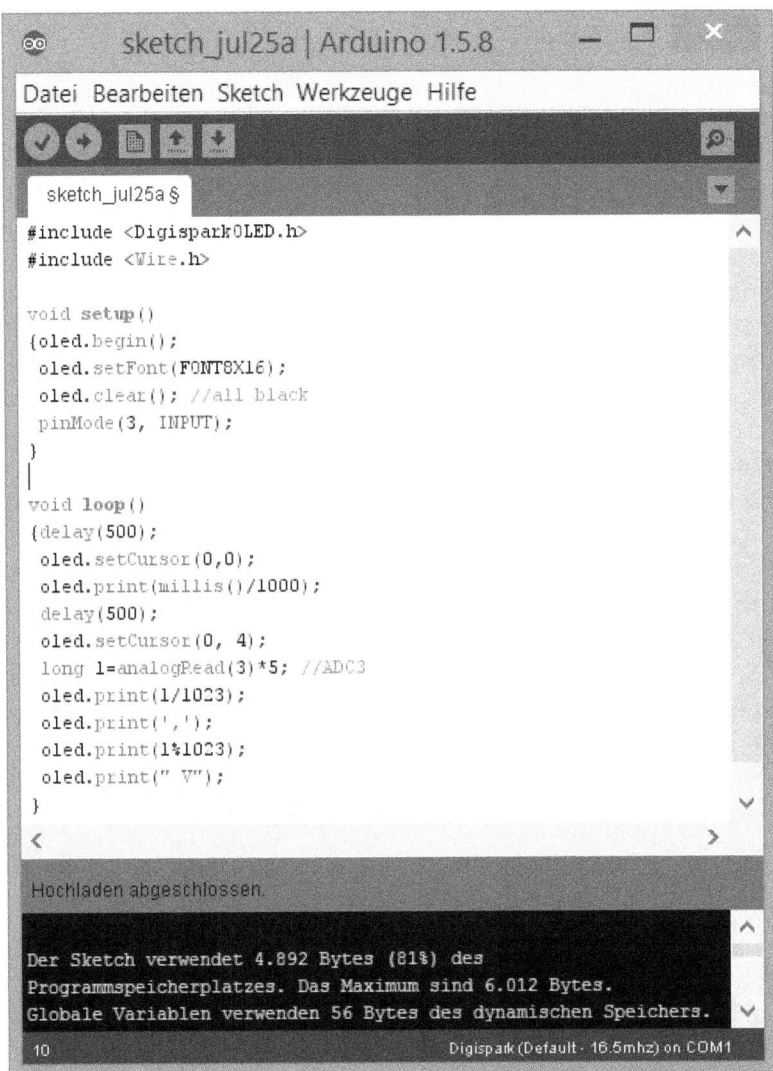

Figure 81: OLED driver using I²C connection

1.4.18 DIGISPARK: REGULATION

To keep something at a desired pre-set value, like a temperature, this is referred to as a regulation. In this case, the current temperature is to be measured and compared to the desired value. In one variation, a controller engages accordingly. This intervention can be carried out continuously and steadily, or discontinuous. The simple discontinuous two-point control is applied here - although a PID Library for the Arduino IDE exists, allowing continuously control - even for memory reasons. The principle of a two-point controller can be found in simple ovens, hot plates or iron. Even many laptop fans only know on and off. It is reciprocated between two points and action switches, wherein the mean is intended to correspond to the desired set point. Since a temperature control system - as described above - behaves as an electric R/C combination, the heating and cooling curves look very similar to the charge and discharge of a capacitor. In a first test, the two-position controller is to be realized without a fan. In this case, there is a desired reference value and two switching thresholds which are respectively above and below the set point. This switching differential is referred to as hysteresis. If a set point value of e.g. 28 °C is selected, the switching thresholds could be at 30 °C (fan on) and at 26 °C (fan off). Thus hysteresis is 4 degrees. The internal temperature sensor of the Digispark serves as the sensor. First only the LED will be switched depending on the threshold and cooling is done manually. On successful preliminary test, a fan will be used.

1.4.19 DIGISPARK: TWO-POINT-REGULATION BLUETOOTH

While developing status information of the control is helpful. Since the direct serial connection to the IDE is not given using a Digispark, initially control is given using Bluetooth/Smartphone. The OLED variant is possible, but is somewhat confusing in the source code at this stage. The code is divided into three main parts: the initialization, the temperature measurement and the main loop. The main loop is the actual controller:

- Temperature measurement
- If higher than upper threshold, then fan on
- If lower than lower threshold, then fan off
- Control outputs
- Wait for some time

In C this loop looks like this:

```
void loop()
{int t= temp();
 if(t>hi)digitalWrite(OUT,HIGH);
 if(t<lo)digitalWrite(OUT,LOW);
 Serial.print(lo);Serial.print("\t");
 Serial.print(t); Serial.print("\t");
 Serial.print(hi);Serial.print("\t");
 Serial.println(digitalRead(OUT));
 delay(1000);
}
```

The measurement of the internal temperature of the Digispark has been used above in the USB keyboard measurement and supplies unchanged the temperature in °C.

```
int temp()
{analogReference(INTERNAL1V1);
 int raw = analogRead(A0+15);
 raw -= 7; // used to calibrate
 int in_c = raw - 273; // celcius
 analogReference(DEFAULT);
 return in_c;
}
```

Using a proper initialization at the beginning, the first Sketch for regulation is created.

```
#include <SoftSerial.h>
#include <TinyPinChange.h>
#define hi 30 //obere Schwelle
#define lo 26 //untere Schwelle
#define RX  2 //BluetoothSerial
#define TX  3 //BluetoothSerial
#define OUT 1 //Lüfter/LED

SoftSerial mySerial(RX, TX);
#define Serial mySerial

void setup()
{Serial.begin(9600); //HC06 mit 9600
 pinMode(OUT, OUTPUT);
```

Using a Bluetooth terminal on a smartphone or tablet temperature values can be observed if a HC06 Bluetooth adapter is connected to pin 2/3 of Digispark. By blowing or waving some paper, the temperature can be manually adjusted; actually this is supposed to take place by a fan. With an old CPU fan, a power bank and a relay now a small fan control can be build using the Digispark. The choice of components is based on their availability. Here, at pin 1 (LED) a 5 Volt relay is controlled by a Digispark. This in turn controls the fan circuit, which is operated with the adjustable power bank depending on the desired fan speed using 5 to 12 Volts. With low power consumption of another fan cooling can be done without a relay to pin 1, but in this case the fan speed was insufficient.

In the right order and the right environment figure 83 arises.

Figure 82: OLED Digispark Relay-Fan

```
#include <SoftSerial.h>
#include <TinyPinChange.h>
#define hi 30    //obere Schwelle
#define lo 26    //untere Schwelle
#define  RX   2 //BluetoothSerial
#define  TX   3 //BluetoothSerial
#define  OUT  1 //Lüfter/LED

SoftSerial mySerial(RX, TX);
#define Serial mySerial

void setup()
{Serial.begin(9600);
 pinMode(1, OUTPUT);
}

int temp()
{analogReference(INTERNAL1V1);
 int raw = analogRead(A0+15);
 raw -= 7; // used to calibrate
 int in_c = raw - 273; // celcius
 analogReference(DEFAULT);
 return in_c;
}

void loop()
{int t= temp();
 if(t>hi)digitalWrite(OUT,HIGH);
 if(t<lo)digitalWrite(OUT,LOW);
 Serial.print(lo);Serial.print("\t");
 Serial.print(t); Serial.print("\t");
 Serial.print(hi);Serial.print("\t");
 Serial.println(digitalRead(OUT));
 delay(1000);
}
```

Übersetzen abgeschlossen.

Figure 83: Regulation -sketch for Digispark in the IDE

1.4.20 DIGISPARK: TWO-POINT-REGULATION OLED

A regulator actually can work without a display, but if a small OLED display is available, the information and possibly changed settings can be displayed.

```
#include <DigisparkOLED.h>
#include <Wire.h>

#define hi 30      //obere Schwelle
#define lo 26      //untere Schwelle

void setup()
{oled.begin();
 oled.setFont(FONT8X16);
 oled.clear();
 pinMode(1, OUTPUT);
}

int temp()
{analogReference(INTERNAL1V1);
 int raw = analogRead(A0+15);
 raw -= 6;
 int in_c = raw - 273; // celcius
 analogReference(DEFAULT);
 return in_c;
}

void loop()
{delay(1000);
 int t= temp();
 oled.setCursor(0, 0);
 oled.print(millis()/1000);
 oled.setCursor(0, 2);
 oled.print(t);oled.print(" C");
 if(t>hi)digitalWrite(1,HIGH);
 if(t<lo)digitalWrite(1,LOW);
 oled.setCursor(0, 4);
 oled.print(lo);oled.print("\t");
 oled.print(t); oled.print("\t");
 oled.print(hi);oled.print("\t");
 oled.println(digitalRead(1));
}
```

1.4.21 DIGISPARK: TWO-POINT-REGULATION ADJUSTABLE

Or "The controlled controller". Depending on the ambient temperature, the test run can be awkward, as no parameters can be changed at runtime. Via Bluetooth, the change of the target value or set point and hysteresis is programmatically easier to solve as rotary switches or buttons on a circuit board or a breadboard.

To get this going there is an inquiry in the main loop whether serial data is available from the outside world. If true and the expected format matches, the corresponding parameter or the corresponding variable is changed. In the controller example, this will be the setpoint and hysteresis. To keep the sketch short, no checking is done, so the user has full responsibility. For clarity, this will take place in a separate routine.

```
void processSerial()
{if(!Serial.available())return;
 if(Serial.find("S"))soll=Serial.parseInt();
 if(Serial.find("H"))hyst=Serial.parseInt();
}
```

If any serial device is sending a character string via Bluetooth starting with an "S", then the set point is set, if there is a "H", the hysteresis is updated. An "S40" sets a target value of 40 °C, a "H2", the hysteresis to 2 degrees. Thus, the thresholds result to 41 °C and 39 °C.

The two constant definitions *lo/hi* are omitted and replaced by *soll/hyst* variables. This results in a modified sketch. Unfortunately, the two methods from SoftSerial *find* and *parseInt* are not implemented, so this easy way fails. The method *readuntil* is not available too. That is why two separate routines are constructed that will remedy the situation. A simple own *readuntil* is to read characters from the interface and copy to an array until a defined end character occurs. A timeout is not provided. The routine behaves similarly to an input in an old BASIC system. The sketch stops until the entry has been made. Parameters are the end character, the array and the length of the array. Return value is the number of characters stored in the array without the end character. The string is null-terminated. The home-made routine *processSerial* checks whether a character is present and if true, a maximum of 20 characters terminated by a line feed are read and analysed with help of *readuntil*. If a correct parameter is detected, the corresponding variable (e.g., *soll*) is set. The method *sscanf* requires surprisingly little space and is used here for convenience as *parseInt* replacement.

```
#include <SoftSerial.h>
#include <TinyPinChange.h>

#define  RX    2 //BluetoothSerial
#define  TX    3 //BluetoothSerial
#define  OUT   1 //Lüfter/LED

SoftSerial mySerial(RX, TX);
#define Serial mySerial

int soll=28;
int hyst=04;

void setup()
{Serial.begin(9600); //HC06 mit 9600
 pinMode(OUT, OUTPUT);
}

void loop()
{processSerial();
 int t= temp();
 int lo=soll-hyst/2;
 int hi=soll+hyst/2;
 if(t>=hi)digitalWrite(OUT,HIGH);
 if(t<=lo)digitalWrite(OUT,LOW);
 Serial.print(lo);Serial.print("\t");
 Serial.print(t); Serial.print("\t");
 Serial.print(hi);Serial.print("\t");
 Serial.println(digitalRead(OUT));
 delay(1000);
}

int temp()
{analogReference(INTERNAL1V1);
 int raw = analogRead(A0+15);
 raw -= 6;
 int in_c = raw - 273; // celcius
 analogReference(DEFAULT);
 return in_c;
}

//------ Erastzroutinen ----------------
```

```
int readuntil (char ct,char *s,  int len)
{int ix=0;char c;
 do
 {c=Serial.read();
  if(c!=-1)s[ix++]=c;
  if (ix>=len)return ix;
 }while(c!=ct);
 if(ix>0)ix--;
 s[ix]=0;
 return ix;
}

void processSerial()
{char s[20];int i;
 if(!Serial.available())return;
 readuntil('\n',s,20);
 if(1==sscanf(s,"S%d",&i))soll=i;
 if(1==sscanf(s,"H%d",&i))hyst=i;
}
```

After start, the sketch shows the following data in one line:

```
Lower treshold
Current value
Upper treshold
Pin 1 Value
```

Expressed in figures, this looks like this:

```
22    20    24    0
```

Here, the lower threshold is 22 °C, the measured temperature 20 °C, and the upper threshold 24 °C. Since the upper threshold has not been reached, the fan connected to pin 1 remains off.

Using a smartphone or tablet connected to Digispark via the *HC06* Bluetooth adapter, the set point can be changed as desired. An input of "S15", with a return, results in setting the target value to 15 °C. A "H2" sets a hysteresis to 2 degrees, so the switching thresholds are calculated one degree above and below the set point:

```
14    20   16    1
```

If the actual value is 20 °C, the controller should now activate the fan at pin 1 as the measured temperature is above the upper threshold of 16 °C. In this way, the influence of the hysteresis can playfully be explored in conjunction with the switching frequency.

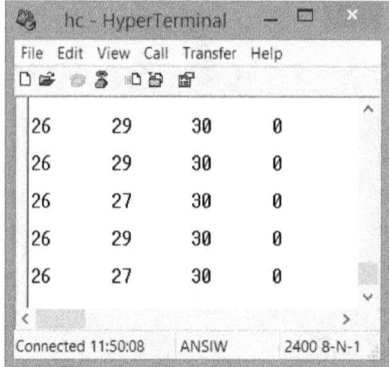

Figure 84: Control control in Hyper-terminal

Here are the pictures of HyperTerminal running Win8.1/32 on a Venue Pro 8 tablet and COM4 with automatic baud rate in conjunction with the Bluetooth adapter *HC06*. The input from a terminal must end with a line feed (LF) to complete; otherwise the routine *readuntil* will never return. The appropriate settings can be found in the English version of Hyper-Terminal *File/Properties/Settings/ASCII Setup*. With the Bluetooth terminal from the Android Play-Store similar screenshots emerge on a smartphone.

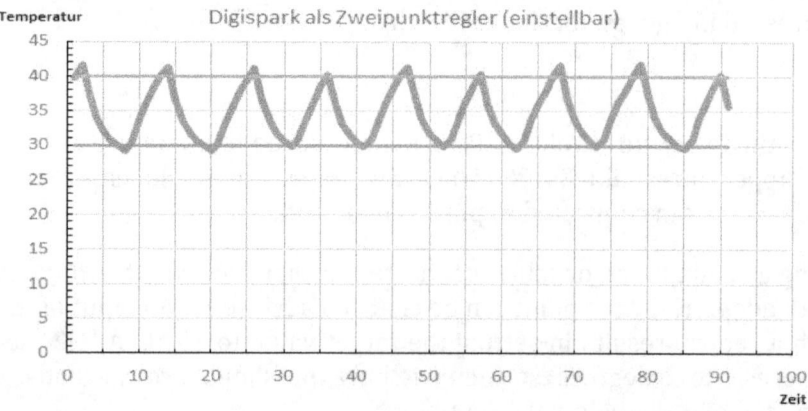

Figure 85: Digispark two point regulations using adjustable hysteresis (Excel chart)

1.4.22 DIGISPARK: MEETS COMPACTDEFINITION

The Digispark is to cooperate with a finished and not for this module intended software. With Compact definition, an own coded Windows executable exists that worked in its time using serial RS232-Interfaces. Then the Arduino became widespread and an old interface was emulated to get the Software working using this microcontroller. The contribution is still on:

http://www.hjberndt.de/soft/ardcompact.html

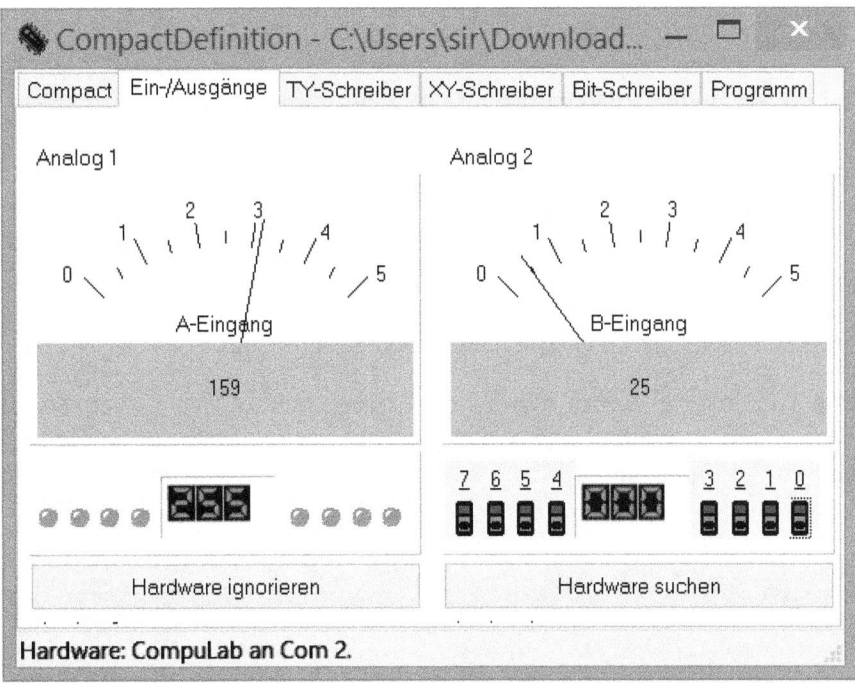

Figure 86: Digispark to CompactDefinition as fake CompuLab at Com2

However a few hurdles are to overcome. The following steps can lead to success.

- Download CompactDefinition 1.75b
  - http://www.hjberndt.de/soft/CompactDefinition%201.75.zip
- Download and adjust the sketch as described below
  - http://www.hjberndt.de/soft/sketch_CLAB_v1.zip

- Install FTDI adapters

Under current Windows versions, a Screen Protect prevents the start of Compact. In Win8.1 a click on details and so on in the warning dialog box starts the harmless software anyway. The *FTDI-adapter* driver search begins when it is first clamped to the USB port, then Windows assigns a COM interface (for example, COM16). In the properties of the serial interface now the lowest possible COM interface (for example, COM2) has to be selected. Explanations are on:

http://www.hjberndt.de/soft/indexcom.htm

The reason is that the hardware search runs slowly in this software version and starts at COM1. It searches sequentially according to a device that responds with 9600 baud to byte 13 with a 2. If this is true, Compact is again fooled to see an Arduino which in turn emulates an old CompuLab. The sketch is adjusted according to Digispark conditions. The SoftSerial interface is created as described in Digispark - Bluetooth keyboard, with RX and TX to pin 2 and pin 1. There are two analogue inputs: PIN 3 and PIN 4. PIN 0 can be used as a digital output.

With two analog inputs two-channel time comparative measurements are possible, as well as an *x/y* plot that detects the direct dependence of the two analogue values. The digital output is for measurement purposes available as a switchable voltage source.

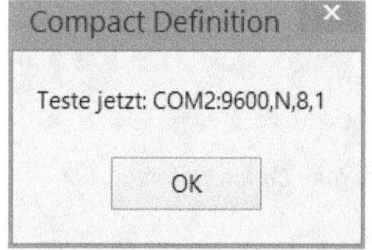
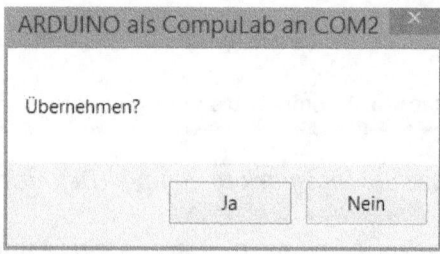

Figure 87: Finding hardware holding down the Ctrl key

```
#include <SoftSerial_INT0.h>
#define P_RX 2
#define P_TX 1
SoftSerial Bluetooth(P_RX, P_TX);
#define Serial Bluetooth
```

```
#define AIN1 60
#define AIN2 58
#define DIN 211
#define DOUT 81

byte Ains[]   = {3,4}; //ANALOGPINS

void setup()
{ Serial.begin(9600);
  pinMode(0, OUTPUT);
  for(int i= 0;i<2;i++)pinMode(Ains[i], INPUT);
}

void loop()
{ int i,val,inbyte ;byte b;
  val = Serial.available(); //Was da?
  if (val>0)
  {inbyte=Serial.read(); //abholen
   delay(1);
   switch(inbyte)
   { case 13  : Serial.write(2);delay(2);break; //ID
     case DIN : Serial.write(255);break; //Für Compact
     case AIN1: Serial.write(analogRead(3)>>2);break;
     case AIN2: Serial.write(analogRead(2)>>2);break;
     case DOUT: b=Serial.read(); //Ausgabebyte holen
                digitalWrite(0,b!=0?HIGH:LOW);
                break;
     default:   break;
   }
  }
  delay(5);
}
```

Controlling the digital output PIN1 and the connected LED would even
be prettier. That is why another library for SoftSerial is used. The analogue inputs now are Pin 4/5, the digital output pin 1 (LED). The first
section is of Digispark - acquired adjustable two-position regulator. The
following lines are mostly unchanged.

```
#include <SoftSerial.h>
#include <TinyPinChange.h>

#define  RX   2 //BluetoothSerial
#define  TX   3 //BluetoothSerial
```

```
#define  OUT   1 //Lüfter/LED

SoftSerial mySerial(RX, TX);
#define Serial mySerial

#define AIN1 60
#define AIN2 58
#define DIN 211
#define DOUT 81

byte Ains[]  = {4,5}; //ANALOGPINS

void setup()
{ Serial.begin(9600);
  pinMode(0, OUTPUT);
  for(int i= 0;i<2;i++)pinMode(Ains[i], INPUT);
}

void loop()
{ int i,val,inbyte ;byte b;
  val = Serial.available(); //Was da?
  if (val>0)
  {inbyte=Serial.read(); //abholen
   delay(1);
   switch(inbyte)
  { case 13   : Serial.write(2);delay(2);break; //ID
    case DIN : Serial.write(255);break; //Für Compact
    case AIN1: Serial.write(analogRead(2)>>2);break;
    case AIN2: Serial.write(analogRead(0)>>2);break;
    case DOUT: b=Serial.read(); //Ausgabebyte holen
               digitalWrite(OUT,b==1?HIGH:LOW);
               break;
    default:   break;
  }
 }
 delay(5);
}
```

If the sketch is uploaded, the RX/TX lines are adjusted and the hardware has been found by compact definition, the LED can be controlled using bit 0.

Because this Windows software has its own simple programming environment, a blinking program without C and without Arduino IDE now is possible.

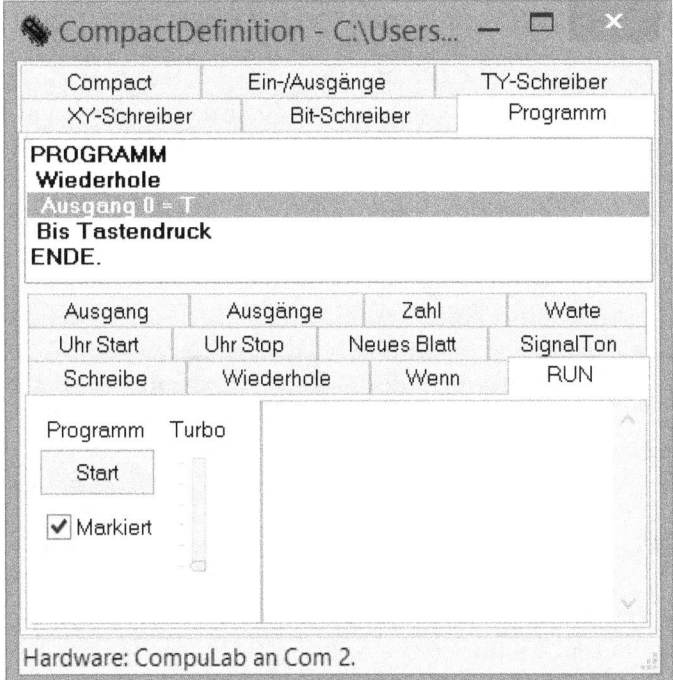

Figure 88: Digispark Blink programmed in Compact Definition

The program runs in a loop and in each pass the state of pin 1 is toggled (T). However, this only works as long as there is a connection.

1.4.23 DIGISPARK: 50 LED RHINETOWERCLOCK (RHEINTURMUHR)

The "Rheinturm" with its decimal clock and the many LEDs can be easily displayed using the small Digispark and a WS2812b LED strip. Since each RGB LED has its own controller, single wire control is possible. The details of the aka Neopixel controller are encapsulated in ready to use libraries for the Arduino that work without any issue for the Digispark. Adafruit provides such a library.

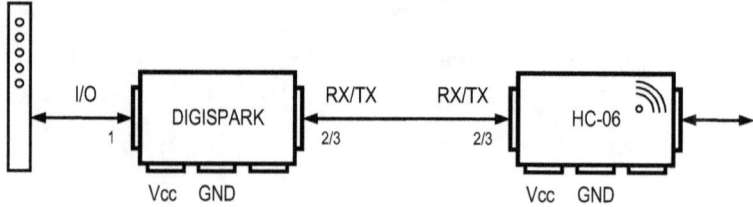

Figure 89: Digispark controls WS2812 with 50 LED as RhineTowerClock

The special feature of this sketch is: it just fits into memory and run well. The clock is slightly inaccurate due to the internal clock frequency while Digispark does its job, but in the last remaining bytes external synchronization could be accommodated using a serial interface. By means of SoftSerial two wires RX/TX are connected to connect to the outside world in various ways: through *ESP8266* and TCP/IP or FTDI-adapter by wire and USB. Next the *HC06*-Bluetooth module is to be used to set the clock or synchronize the hardware.

Because of little space the usual *int* variables in other listings for Rhine Tower on *hjberndt.de* are reduced to byte variables, resulting in the *Serial.print* with a strange behaviour. The sketch sends the time string *hh:mm:ss* every second to serial and in exact that format the clock can be set in the opposite direction. Once an incoming serial character is available the sketch stops and waits for the return key in the form of a line feed (enter:,\n'). Due to lack of memory the entry is not checked. Entering *12:34:56* will set the time.

```
#include <SoftSerial.h>
#include <TinyPinChange.h>
#include <Adafruit_NeoPixel.h>
#define PIN    1
#define   RX    2 //BluetoothSerial
#define   TX    3 //BluetoothSerial
```

```
#define NUMPIXELS      50
#define ON     Color(31,31,31) // 120 mA max for all
#define OFF    Color(1,0,0)
#define MARK   Color(15,0,0)

//LEDS SEC MIN HOUR
byte oneSecond[] = {0,1,2,3,4,5,6,7,8};              //09
byte tenSecond[] = {10,11,12,13,14};                 //15
byte oneMinute[] = {16,17,18,19,20,21,22,23,24};     //25
byte tenMinute[] = {26,27,28,29,30};                 //31
byte oneHour[]   = {32,33,34,35,36,37,38,39,40};     //41
byte tenHour[]   = {42,43};                          //44
byte Marker[]    = {9,15,25,31,41,44,45,46};
byte s10=0,s1=0; //ONESECOND TENSECOND
byte m10=5,m1=9;
byte h10=2,h1=3;

Adafruit_NeoPixel pixels = Adafruit_NeoPixel(NUMPIXELS,
PIN, NEO_GRB + NEO_KHZ800);

#define  RX   2 //BluetoothSerial
#define  TX   3 //BluetoothSerial
SoftSerial mySerial(RX, TX);
#define Serial mySerial

void setup()
{pixels.begin();
 // This initializes the NeoPixel library.
 for(int i=0; i<sizeof(Marker); i++) pixels.setPixelColor
(Marker[i], pixels.MARK);
 Serial.begin(9600);
}

void loop()
{static uint32_t prevMillis = 0;
 if (Serial.available())processSyncMessage();
 while (millis() - prevMillis >= 1000)
 {prevMillis += 1000;
  s1++;                          //clock set       //s++;
  if(s1>=10){s1=0;s10++;}
  if(s10>=6){s10=0;m1++;}        //m1++;
  if(m1>=10){m1=0;m10++;}
  if(m10>=6){m10=0;h1++;}        //h1++;
  if(h1>=10){h1=0;h10++;}
  if(h10>=2 && h1>=4){h10=0; h1=0;}
  digitalClockDisplay();
```

```
   digitalRhineTower();
  }
}

void digitalRhineTower()
{int i;
 for(i=0;i<sizeof(oneSecond);i++)
   (s1<=i?pixels.setPixelColor(oneSecond[i], pix-
els.OFF):pixels.setPixelColor(oneSecond[i], pixels.ON));
  for(i=0;i<sizeof(tenSecond);i++)
   (s10<=i?pixels.setPixelColor(tenSecond[i], pix-
els.OFF):pixels.setPixelColor(tenSecond[i], pixels.ON));
  for(i=0;i<sizeof(oneMinute);i++)
   (m1<=i?pixels.setPixelColor(oneMinute[i], pix-
els.OFF):pixels.setPixelColor(oneMinute[i], pixels.ON));
  for(i=0;i<sizeof(tenMinute);i++)
   (m10<=i?pixels.setPixelColor(tenMinute[i], pix-
els.OFF):pixels.setPixelColor(tenMinute[i], pixels.ON));
  for(i=0;i<sizeof(oneHour);i++)
   (h1<=i?pixels.setPixelColor(oneHour[i], pix-
els.OFF):pixels.setPixelColor(oneHour[i], pixels.ON));
  for(i=0;i<sizeof(tenHour);i++)
   (h10<=i?pixels.setPixelColor(tenHour[i], pix-
els.OFF):pixels.setPixelColor(tenHour[i], pixels.ON));
  pixels.show();
}

void digitalClockDisplay()
{Serial.print(h10+0); //...digispark spezifisch?
 Serial.print(h1+0);
 Serial.print(":");
 Serial.print(m10+0);
 Serial.print(m1+0);
 Serial.print(":");
 Serial.print(s10+0);
 Serial.println(s1+0);
}

//------ serial Hilfsroutinen ----------
int readuntil (char ct,char *s, int len)
{int ix=0;char c;
 do
 {c=Serial.read();
  if(c!=-1)s[ix++]=c;
  if(ix>=len)return ix;
 }while(c!=ct);
```

```
 if(ix>0)ix--;
 s[ix]=0;
 return ix;
}

void processSyncMessage()
{char t[20];
 readuntil('\n',t,20);
 //if(3==sscanf(t,"%02d:%02d:02d",&h,&m,&s))
 //TOO BIG
 s1  = t[7]-'0';
 s10 = t[6]-'0';
 m1  = t[4]-'0';
 m10 = t[3]-'0';
 h1  = t[1]-'0';
 h10 = t[0]-'0';
}
```

Compiled using Digispark IDE 1.5.8

A 7-line VBScript synchronizes the clock with the system time from a Windows tablet, if the *HC06* module is available as COM4 every 15 seconds like this:

```
Set objFSO = CreateObject("Scripting.FileSystemObject")
Set f = objFSO.OpenTextFile("COM4:", 8, True)
While true
       WScript.Sleep 15000
       f.WriteLine(time)
wend
f.Close
```

Figure 90:
Time2com4.vbs on a Desktop synchronizing Digispark time

Once started, this script runs in the background and communicates the time to the Digispark. It is visible at runtime in the Task Manager only. If the interface is not responding, an error message will occur.

If an Android device is preferred, then the corresponding 7 *rfo*-Basic lines are:

```
INCLUDE btopen.bas
DO
 TIME hour$, Month$, Day$, Hour$, Minute$, Second$
 A$=hour$+":"+minute$+":"+second$
 BT.WRITE a$
 PAUSE 5000
UNTIL false
```

The opening of the Bluetooth interface is *btopen.bas* in the file and corresponds to the listing in [2]. The content is specified in the last Chapter in this book too. There *Aeronautical Radio Time Announcement* or *Flugfunkzeitansage* via DCF39 is a variant of the Rhine Tower to synchronize in a special manner. This wayward method of control is listed in the last part of that chapter.

2 Software Elements

There are many special applications out there that can handle serial data of measured values, but none of them can handle all those strange things that a user wants to do or solve. For this reason there is a chapter 2. Here are short briefings of some software elements possibly representing a kind of toolbox. Depending on the problem that has to be solved, the one or the other element on Android and/or Windows might be the missing bit.

2.1 VBS – Visual BASIC Script

In Windows there lives a built-in interpreter, which can be coded like Visual BASIC (VB). This results in a few or more possibilities within the context of this book. While VB is more powerful, VBS can handle files and the execution of software - often a nice to have. Particularly interesting might be the fact that Windows serial RS232 interface is treated as a file since win32. So VBScript is able to talk to FTDI/RS232Adapters and is able to communicate over Bluetooth, however sometimes Windows has some issues using Bluetooth - on this machine.

As a block VBScript could be represented like this:

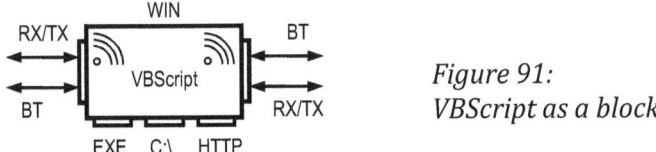

Figure 91:
VBScript as a block

Some BASIC programming skills are required. Teaching this language is not the subject of this essay.

2.1.1 VBS: Coding

The code can be written in any editor (i.e. Notepad). If VB is not completely new, the following syntax might appear familiar:

MsgBox "Hello World".

By renaming the file extension, Windows runs these lines containing the instructions. An additional compiler or interpreter (VB/VBA) for execution is not required. This Interpreter ships - unnoticed by many - as Windows Script Host from Microsoft. This is in fact a batch processor in 'VB'. Some people call it scripting. The professional should turn to Windows PowerShell, but to solve little problems the old CMD.EXE will do the job fine.

To distinguish the various output modes, here a tutorial in brief.

First program "Hello world" ...

- Start editor (for example Desktop, right mouse button)
- Enter *MsgBox "Hello world, 66"*
- Save file and rename to *hallo.vbs* (icon change)
- Doubleclick *hallo.vbs* shows the result

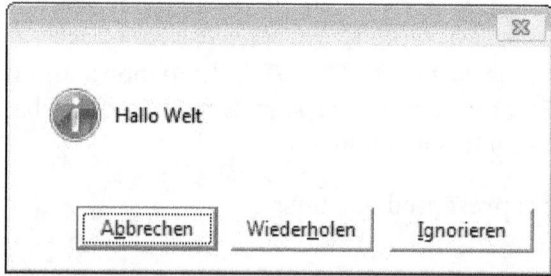

Figure 92:
VBScript output 1

Second program "Hello world"...

- Start editor
- Enter *wscript.Echo "Hello World"*
- Save file and rename it to *hallo.vbs* (icon change)
- Doubleclick *hallo.v*bs and the result is shown

Figure 93:
VBScript output 2

The second program without a message box for output

- Start editor and type *cscript hallo.vbs*
- Save file and rename it to *hallo.bat* (icon change)
- Double click on *hallo.bat,* Windows shows result very briefly
- In *hallo.vbs* append the line *Wscript.sleep 10000*

Figure 95:
VBScript output 3

Now the result is visible for about 10 seconds. A screenshot is shown on the left from a Windows7 system. Using the *WScript* output method might result in many message boxes that all want to be confirmed. Finally, if the file *hallo.bat* contains *wscript hallo.vbs* the fourth variant is obtained: A message box on a 'DOS' window.

2.1.2 VBS: KNOW-HOW, HELP, EXAMPLES

More special scripts require manipulation of ready to run examples and an interactive help. The tool *VbsEdit 7394 by Adersoft* is what a beginner needs. The unregistered evaluation version aims precisely the needs as the many samples are perfectly suitable for testing. Some example only runs when the environment was launched as an administrator. The user should know what is going to happen, by running a VBScript. Many scripts for solutions presented in this book are based on examples from *Samples/Scripting Techniques/Text Files.*

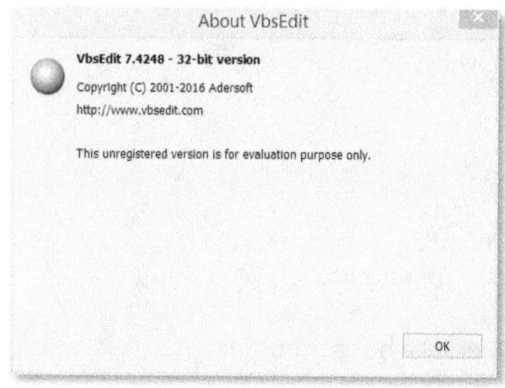

Figure 96: The free version of VbsEdit was used to realize a few things

2.1.3   VBS: DIGISPARK VIA BLUETOOTH

This example demonstrates VBS communication using a Digispark via Bluetooth and the *HC06* Bluetooth module connected to pins 3/4 sending time data (see section "Hardware Elements"), the original example *Writing Data to a Text File* serves as a template. With appropriate changes the Digispark messages will appear in the output window. The script opens COM4 (*HC06*) and reads incoming data in a continuous loop to display them using *WScipt.Echo* in the output window.

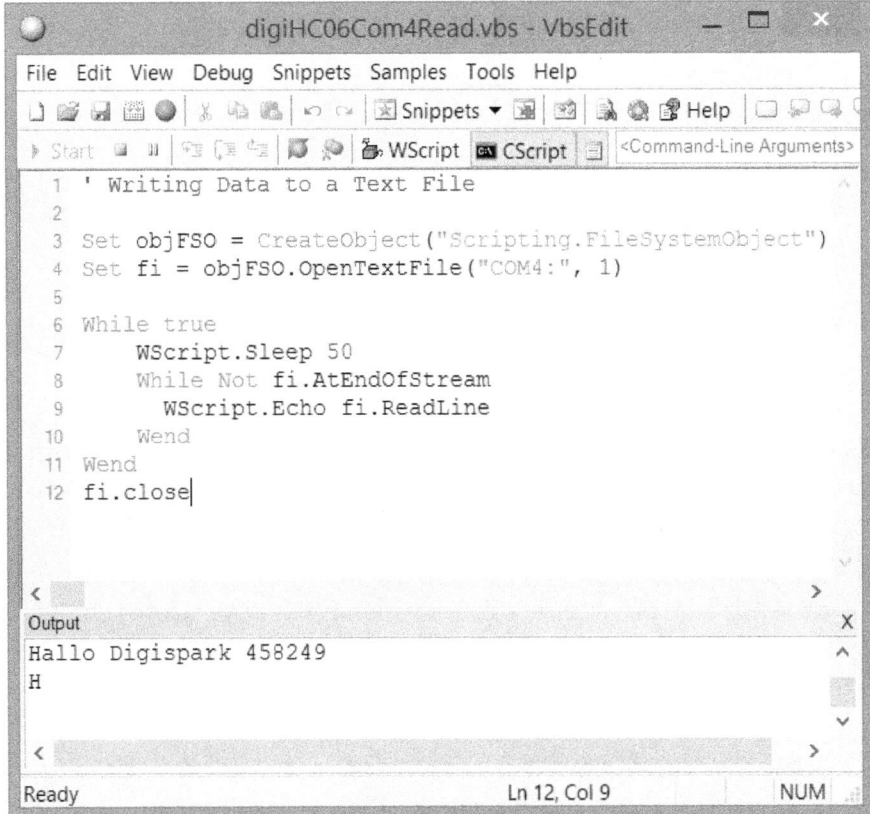

Figure 97: Digispark via Bluetooth using VBScript

```vbs
' Writing Data to a Text File

Set objFSO = CreateObject("Scripting.FileSystemObject")
Set fi = objFSO.OpenTextFile("COM4:", 1)

While true
      WScript.Sleep 50
      While Not fi.AtEndOfStream
            WScript.Echo fi.ReadLine
      Wend
Wend
fi.close
```

2.1.4   *VBS: DIGISPARK WIRELESS TO EXCEL*

This application shows another modified example from *Samples/Microsoft Office*. The script launches EXCEL and inserts values into a cell.

```
' Add Data to a Spreadsheet Cell

Set objExcel = CreateObject("Excel.Application")

objExcel.Visible = True
objExcel.Workbooks.Add
objExcel.Cells(1, 1).Value = "Test value"
```

Combining the two scripts might look like this, returning the result shown. The Digispark as a USB keyboard did the same thing shown in the hardware section, now it is accomplished in a completely different manner.

```
Set objFSO = CreateObject("Scripting.FileSystemObject")
Set fi = objFSO.OpenTextFile("COM4:", 1)
Set objExcel = CreateObject("Excel.Application")

objExcel.Visible = True
objExcel.Workbooks.Add

While true
      WScript.Sleep 50
      While Not fi.AtEndOfStream
            objExcel.Cells(1, 1).Value = fi.ReadLine
      Wend
Wend
fi.close
```

There are only 12 lines of code to get the desired output. Copied to Notepad and saved and renamed (*.vbs) will allow a double click to run the script above. The following things are done in sequence:

- The interpreter executes the script
- COM4 connects (slowly) to the Bluetooth adapter *HC06*, the LED is continuously on if connected. This might fail sometimes, but with a few attempts it finally works
- EXCEL starts up with a blank spread sheet and after a delay the Digispark message appears in cell A1

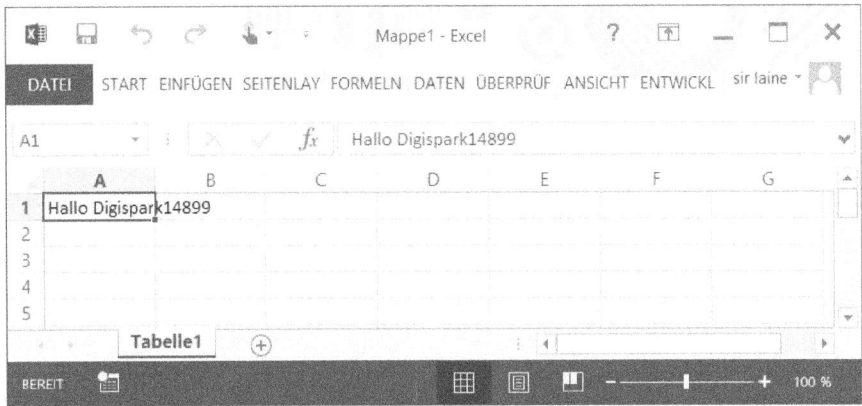

Figure 98:
Digispark provides data via Bluetooth and VBS directly to EXCEL

After another delay the item changes as more data arrive. On errors (error messages may appear at the next attempt) Excel and the Microsoft Scripting Host should be stopped using the Task Manager.

With minor changes the script can handle analogue measurement, automatically transferred from Digispark using VBS - and brought into EXCEL for displaying. The data does not run in sync, due to file buffering. While data is arriving, Excel can do other things i.e. a presentation as a simple chart can be done.

The Script fills the first 10 rows continuous by a loop with data, the chart updates as data arrives. The incoming values are measurements of the 1.1 volt reference to ground (see LM35 Temperature).

```
Set objFSO = CreateObject("Scripting.FileSystemObject")
Set fi = objFSO.OpenTextFile("COM4:", 1)
Set objExcel = CreateObject("Excel.Application")

objExcel.Visible = True
objExcel.Workbooks.Add
ix=1
While true
     WScript.Sleep 50
     While Not fi.AtEndOfStream
```

```
                    objExcel.Cells(ix, 1).Value =
fi.ReadLine
          ix=ix+1
          if ix>10 then ix=1
     Wend
Wend
fi.close
```

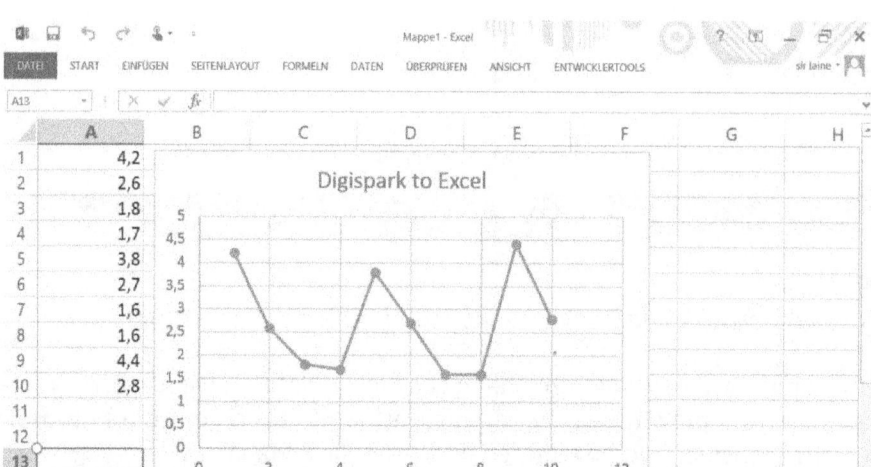

Figure 99: Digispark updates Excel Chart via Bluetooth

Digispark can send data by wire to Excel too. This involves the exchange of the *HC06-* by the *FTDI- adapter.* Via USB (Host)-cable and the appropriate other COM number it can be can accessed in VBS. A touch of the book "Measuring Control and Regulation using Word and Excel" [3] is softly breezing straight through the room.

2.1.5 VBS: TIME TO COM2:

In section Digispark meets *CompactDefinition* the USB-FTDI-adapter was listed as COM2 in the device manager. A short *VBscript* will now transfer data - the current time - from a Windows tablet to a device connected to this serial port. The "file" is opened for appending and at a 5 second interval the time is sent as text with additional display in the output window.

```
Set objFSO = CreateObject("Scripting.FileSystemObject")
Set f = objFSO.OpenTextFile("COM2:", 8, True)
While true
      WScript.Sleep 5000
      f.WriteLine(time):WScript.Echo time
wend
f.Close
```

An extended example could transfer the timestamp to a smartphone by connecting the *FTDI* to a *HC06*-Bluetooth adapter. The smartphone then can receive data via Bluetooth, as shown elsewhere in this book. In a further step these lines could be inserted - without the echo output - to *Notepad.exe* and stored as *time2Com2.vbs* to start the transfer by double-clicking. Using the comments on *NetCat* again elsewhere in this book, it is possible to start and stop this script even from a smartphone.

2.1.6 VBS: INTERNET DATA QUERY "CQ DX"

Imagine a situation where an automated query of data from e.g. an Internet database is required. In this example it is obtaining information about a radio amateur on his call. Of most interest is his location. This solution is used in a larger context in the third part of this book, where the aim is to locate the signal strength of a radio amateur station with low power in the HF range in JT65-transmission technology.

At *http://qrzcq.com/* a call sign query can be done by typing into an input field. The results appear on a correspondingly generated HTML page.

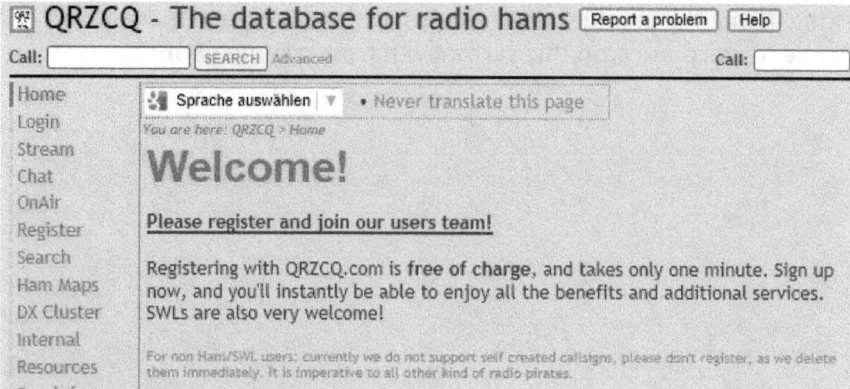

Figure 100: Homepage and found result for a random call sign ZS6WB from the database on the net

A look at the source code finds the desired data starting at line 201, surrounded by a lot of HTML garbage. In [1] three examples are shown of how to get this data by means of the Android *rfo*-Basic and HTML.GET, but here the search is done by *VBScript* and Windows.

The actual call in the script is in the first line, which provides the information found, outputs it, if anything was found. If the call sign is re-turned unchanged, the search failed. Previously the search behavior is observed in a browser and then is determined if a direct call to

http://qrzcq.com/call/ZS6WB

using the address input of the browser provides all the desired data too. If this works sucessfully, the search can proceed without keyboard input.

Figure 101: HTML source code of a search result

After creating the HTTP object the GET call is done. The generated page is provided by *ResponseText* and stored in the string *s*. The search *InStr* finds the position of the specified constant *find* (because of quotes BASIC syntax is somewhat confusing) *<p class=""haminfoaddress""><b style=""text-shadow: 0px 1px 0px #f1f1f1, 0px 1px 3px #999; "">*. The de-sired information is found in the next 100 characters, the country then is found in the fourth array entry after the SPLIT call.

```
WScript.echo GetCallsign("ZS6WB")

Function GetCallsign(sign)
  Const find="<p class=""haminfoaddress""><b style=""text-
shadow: 0px 1px 0px #f1f1f1, 0px 1px 3px #999; "">"
  GetCallsign=sign
```

```
strURL = "http://qrzcq.com/call/"+sign
Set objHTTP = CreateObject( "WinHttp.WinHttpRequest.5.1"
)
 objHTTP.Open "GET", strURL: objHTTP.Send
 s= objHTTP.ResponseText:   ' WScript.Echo Len(s)
 if objHTTP.Status = 200 Then
    ix=InStr(s,find)
    s=Mid(s,ix+Len(find),100)
    sa=Split(s,"<")
    GetCallsign = Mid(sa(4),6)'Country
    'GetCallsign = Mid(sa(3),6)'City
    'GetCallsign = sa(0)'Name
 End if
 Set objHTTP = Nothing
End Function
```

2.1.7 VBS: SPEECH

The speech output in VBS is just as simple as in *rfo*-Basic on Android. If speech is set to English, a sentence of the early computing days should be heard.

```
Set Sapi = Wscript.CreateObject("SAPI.SpVoice")
Sapi.speak "Hi, I'm Eliza. Tell me about your prob-
lem."
```

For a spoken timestamp, as found in [1] the minimal form may look like this:

```
Set Sapi = Wscript.CreateObject("SAPI.SpVoice")
Sapi.speak time
```

The result (Microsoft David) just after breakfast is "*ten hours, twenty minutes and three seconds*", if German voice output is enabled, the whole thing sounds like this: "*Zehn Uhr Zwanzig und drei Sekunden.*"

2.1.8 VBS: RUN/EXECUTE

In VB-Script there is a powerful execute command. Executables can be called directly or by reference by a linked file. If *kmz*-files are mapped to Google Earth, a call to *abc.kmz* will start Google-Earth and loading the kmz-file. Everything works like double-clicking on desktop-icons. Here a small example of another script being executed every minute from a script. It is a kind of VBScript egg timer.

The time announcement by speech was shown some lines above in the previous section. Using the two lines as content of a text file with the filename *time2speech.vbs*, on the desktop, a double click will tell the time. A second script calls this script as a time announcement once per minute (60,000 ms). All this runs to get a well cooked egg for seven minutes as a pure example and placeholder for more meaningful timed tasks.

```
set WshShell = WScript.CreateObject("WScript.Shell")
for i = 1 to 7
        WshShell.Run "time2speech.vbs"
        wscript.sleep 60000
next
wscript.echo "Sieben Minuten sind rum!"
```

These lines can be saved on the desktop too as a **.vbs* file, a double click will run the egg timer. There is an interesting behaviour when two or more speeches overlap. Windows waits a short while for the end of one voice before starting the next, so nothing gets lost - it is latent only - nothing special using windows.

Endless Execution

Software which is closed by mistake and has to start again immediately, the following procedure is conceivable. A batch file named *0.bat* with the content *cscript 0.vbs* calls the script *0.vbs* on execution and directs its output to the DOS window and not to Windows boxes. The VB script contains the following lines:

```
Set WshShell = WScript.CreateObject("WScript.Shell")

While true
        Return = WshShell.Run("ABC.exe", 1, true)
Wend
```

The set parameters of the run function cause a fictitious application ABC.exe to execute, then the function waits until it has finished. An endless loop repeats this sequence endlessly.

The effect can be tested replacing ABC.exe with the Windows editor Notepad.exe. Once the program is started via *0.bat* even after a quit it will come right back, because it is run again and again by the script when it ends. Only closing the batch window stops this loop. If calling the script without *0.bat*, there is no visible process window, the script still remains the same, but the loop is hardly to stop.

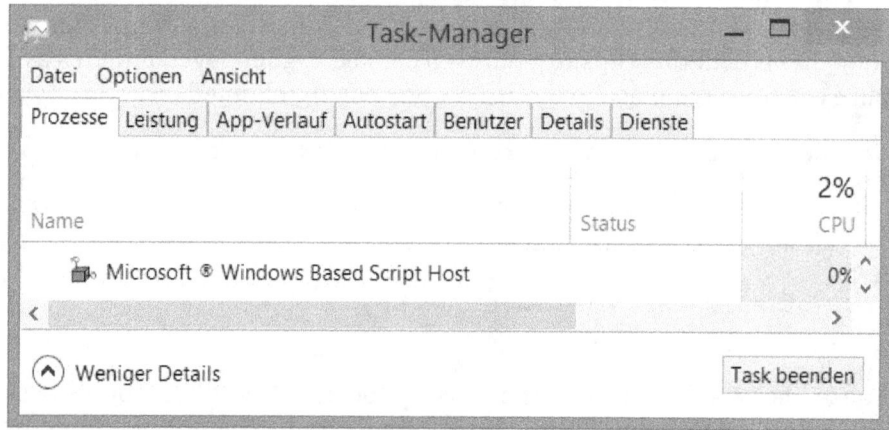

Figure 102: Script Host in Win8.1-Task Manager

It runs until a *taskkill* of the Scripting Host is done by the Task Manager. And of course a Windows reboot ends up this game.

2.1.9 VBS: *BEEP AND MUSIC*

A sound or whole files of music can be simply "spoken" in *VBScript*, if the filepath is correct.

```
Dim objFile
Set objFile = CreateObject("SAPI.SpFileStream.1")
objFile.Open "C:\Windows\Media\tada.wav"
CreateObject("SAPI.SpVoice").Speakstream objFile
Set objFile = nothing
```

2.1.10 VBS: KEYBOARD CONTROL

Windows-compatible programs can usually be operated and controlled using the keyboard. Beside the familiar keyboard shortcuts *Ctrl+C, Ctrl+V* for copying and pasting in applications and Windows can terminated itself using *Alt+F4*. In menu items, the corresponding abbreviations often are given. With the Alt key (pressed 2x) Windows shows keyboard shortcuts in some way. In WordPad - the standard program - thus an "Undo" in addition to the standard *Ctrl+Z* has an equivalent in *Alt+2*.

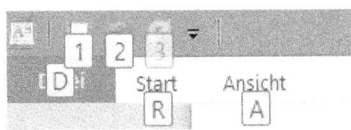

Figure 103: Pressing the Alt key twice to shows keyboard shortcuts of Wordpad

VBS also provides the *SendKeys* method, as it was or is known in VBA, however, without the latest restrictions of VBA (yet). The keyword *SendKeys* in Help of *VbsEdit* explains the way of how to use and shows the calculator-example from Microsoft VBA-Help, which, however, runs here in the *VbsEdit* environment a little off track. To illustrate another example which should run on any Windows system, the Windows Notepad named *Notepad.exe* is used again.

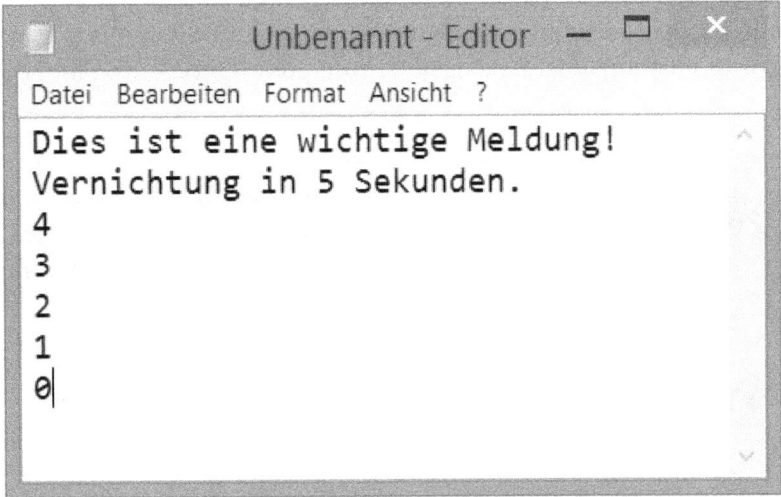

Figure 104: A message box via Notepad showing a countdown

The executable is run by the *Run* statement and should then be the top-most window. A script now sends keystrokes to the window, according the variables *a* and *b*. Thereafter, an output of the countdown is written by a loop and finally the process is terminated using *Alt+F4*. The subsequent storage request is cancelled by moving the focus to the next dialog element via tab. The space-key then pushes the focus element, which cancels storing and closes the program.

```
set WshShell = WScript.CreateObject("WScript.Shell")
WshShell.Run "Notepad.exe"
WScript.Sleep 100
WshShell.AppActivate "Editor"
WScript.Sleep 100
a="Dies ist eine wichtige Meldung!"
b="~Vernichtung in 5 Sekunden.~"
Beenden="%{F4}"
NichtSpeichern="{TAB} "

WshShell.SendKeys a+b
for i = 4 to 0 step -1
 WshShell.SendKeys cstr(i)+"~"
 WScript.Sleep 1000
next
WshShell.SendKeys Beenden+NichtSpeichern
```

Note: All keystrokes get to the active window. By intervening change in focus by the user or perhaps even by a Digispark as a USB keyboard, inadvertently applications can be closed or run, possibly resulting in data loss.

This procedure could be used to control applications which do not use official interfaces such as DDE or COM. However, the application should not ignore Windows programming for whatever proprietary reasons. Using *NetCat* short messages can be displayed initiated by a smartphone or tablet.

2.2 NETCAT

Controlling Windows from an Android phone? Yes, *NetCat* can.

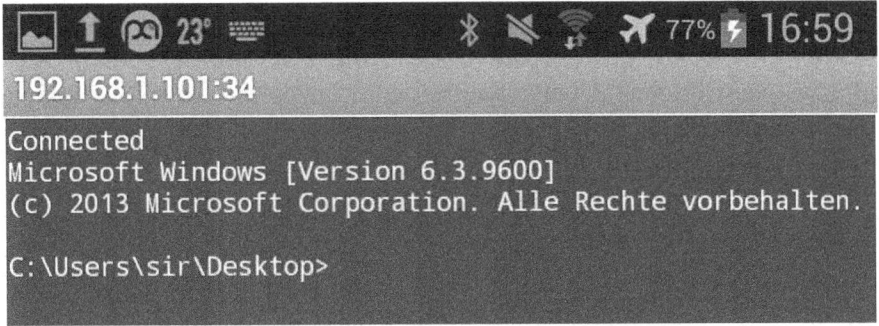

Figure 105: Windows 8.1 on an Android phone

The NetCat tool is software that is usually used in the low level of a command line. Both Linux and Windows systems know this tool. *NetCat* or *nc* can redirect TCP/IP data streams unchanged, resulting in some interesting aspects. Simple client/server applications are easy to create on the fly. Following these possibilities, controlling via TCP/IP can be done and even an Android smartphone can take control of a Windows PC. In a leaflet from the *Internet Netcat Cheat Sheet* by Ed Skoudis the possibilities are summarized. The following sections describe some of these applications or possibilities; they will be illustrated in the context of this book title with some examples. The basic principle of the client is *nc [destination IP] [port]* for establishing a connection, the server or the listener *nc -l -p [local port],* where *nc* is the call to nc.exe or NetCat and the space-separated data or the arguments are the transfer parameters.

2.2.1 NETCAT: FILE COPY TCP/IP

First, a file from PC8 to PC9 is to be copied using TCP/IP and NetCat. Either Windows computers or tablets are equipped with NetCat living as *nc.exe* in each directory *c:\temp\nc* on both PC. As working directory *c:\temp* is used. For security the path of NetCat has not been included in the path variable that is why always the entire path has to be specified, i.e. *c:\temp\nc\nc.exe*, when called.

All commands for a command line or the command prompt can also be passed to a text file if the extension of the text file is *.bat*, a so called batch file. The file to copy is located in the *c:\temp* directory of PC8 and its filename is *hallopc9.txt* containing some text *"Hello NetCat on PC9."* This file has to be copied to PC9 in that local Temp directory with the filename *pc8.txt*.

Receiver

The receiving PC9 listens to port 59 and NetCat directs all input from there to the file *pc8.txt*. A batch file or the command line then contains the following characters:

```
c:\temp\nc\nc -l -p 59 > pc8.txt
```

When this line executes, port 59 is obeyed and the incoming characters are directed to the file *pc8.txt*. Immediately at execution NetCat creates the output file in the directory (temp). The file is still empty, with a length of 0 bytes. NetCat waits for a connection by PC8 as sender.

Transmitter

The sending PC8 finds *hallopc9.txt* - the file to be transferred - in the local Temp directory. A batch file named *NcSendFile.bat* or the command line contains the following string:

```
C:\temp\nc\nc -w3 192.168.1.101 59 < hallopc9.txt
```

The first part is the call to NetCat on this PC8, the *w3* parameter lets *nc* wait three seconds for a connection on port 59 to PC-9 and its local IP as next argument. As soon as a connection is established the content of the file *hallopc9.txt* is sent. As a result, there is a file *pc8.txt* with the contents of the file hallopc9.txt from the Temp directory of PC8 in the Temp directory of PC9. The file length of *pc8.txt* has changed, because its content is "*Hello NetCat on PC9*".

2.2.2 NETCAT: RELAY

The short-term *listener-to-client relay* from the cheat sheet is to be applied in this environment. This requires a router equipped both PC with corresponding IP addresses. If using a mobile router and a SSID ES830-C813, the addresses could have the following form:

PC9:   192.168.1.100

PC8:   192.168.1.101

In the original cheat sheet reads:

```
C:\> echo nc [TargetIP] [Port] > relay.bat
C:\> nc -l -p [LocalPort] -e relay.bat
```

The first line creates a file named relay.bat and the content gets the call to NetCat and the destination IP as parameters. Without PATH variable, this means for the chosen configuration:

```
echo c:\temp\nc\nc 192.168.1.101 58 > relay.bat
```

After execution, there is a new file *relay.bat* in the working directory (temp) with the text content "*c:\temp\nc\nc 192.168.1.101 58*".

The second line ensures that PC9 is listening on port 59 and waits for a connection, then executing *relay.bat*, which in turn calls *nc* to redirect all inputs to port 58 on PC8 using IP 192.178.1.101. Specifically, the line for this environment is:

```
c:\temp\nc\nc -l -p 59 - e relay.bat
```

Executing this line on PC9 in the *cmd*-window, corresponding content appears. Now *NetCat* is waiting for PC9 on port 59 to connect.

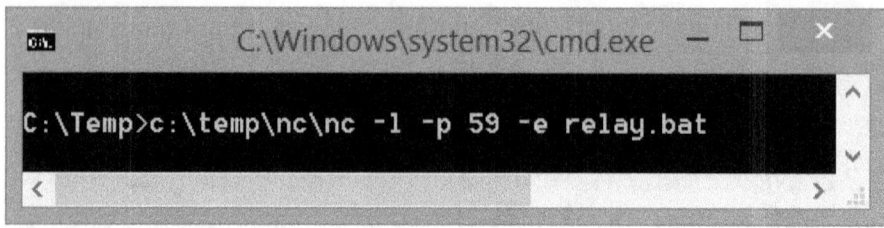

Figure 106: NetCat as Relay

Now HyperTerminal is running on PC8 using the connection parameters for PC9

Hostaddress: 192.168.1.100
Port Number: 58
Connect using: TCP/IP

When the connection is established, the line

```
C:\Temp>c:\temp\nc\nc 192.168.1.101 58
```

appears in the terminal window. *Relay.bat* has been executed on PC9, which redirects all output of PC9 via *nc* to the specified port of PC8.

2.2.3 NETCAT: CHAT

A chat application in one line is possible using NetCat on Windows in the minimalistic form:

```
nc -L -p 35
```

NetCat is waiting for a connection on port 35 and then redirects all standard entries. All keystrokes in a "DOS window" get to the recipient via TCP/IP. In the above assumptions, a batch file with the content *C:\Temp\nc\nc L -p 35* with a filename *NcLp35.bat* can be stored on the desktop. After double clicking the file, NetCat is listening on port 35.

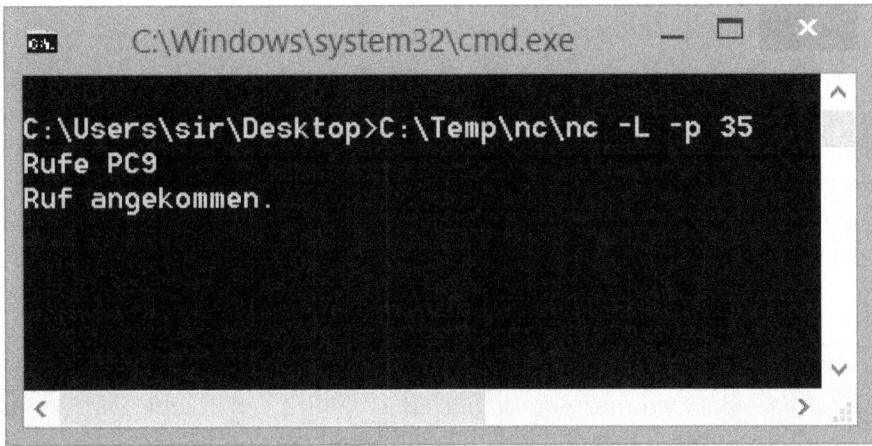

Figure 107: NetCat listening at Port 35

The Android smartphone or tablet calls Port 35 by means of a TCP/IP-client on the same local network 192.168.1.101 as PC9. As soon as both devices are connected the chat can begin. Since the capital letter *L* is used, NetCat is listening to the client even after an interruption.

Figure 108: Android smartphone and "PC9" combined to create a chat on the net

2.2.4 NETCAT: EXCECUTE ON CALL

In the third Chapter *NetCat* is used to control Windows from other devices connected via TCP/IP. The most important *NetCat* parameter in this context is the *-e* execute. In the previous section, this parameter has been used to start *relay.bat*. Starting the command processor *cmd.exe*, you end up in a kind of DOS world with the license to control Windows. If you start a batch file with the text content

```
c:\temp\nc\nc -L -p 34 -e cmd.exe
```

on a Windows tablet or PC, *NetCat* listens on port 34 on the Windows computer. When connected, all output is redirected to the caller and *cmd.exe* is started. Thus, the caller has a lot of control and management options.

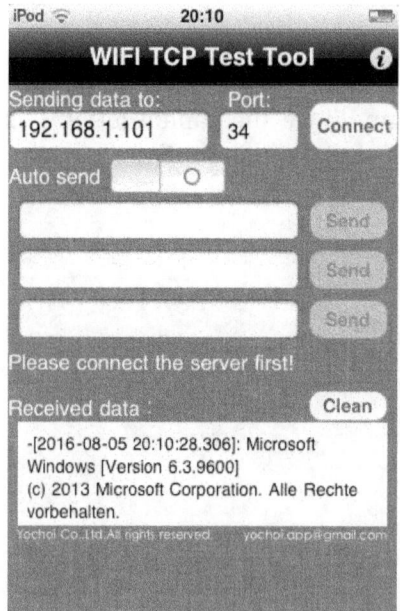

Figure 109:
iPod Touch 2G and Win8.1

An Android smartphone using the app TCP/IP-Terminal from the Play Store can initiate the call as a client on port 34 via the IP of the Win tablet, and then the screenshot shown at the beginning of this section will appear: Windows 8.1 in Android-control.

Android runs notepad.exe

Android launches notepad.exe, or tries some of the known commands from the early days of DOS computing - tree, manoeuvre with cd through directories. Calling ipconfig to get info's or tasklist/taskkill to list and stop running applications. To run the standard editor, entering

```
notepad.exe
```

on the smartphone will do. In addition some arguments can open files (depending on user rights),

```
notepad.exe c:\windows\win.ini
```

shows an old Windows file that has been replaced by the registry.

Android kills notepad.exe

To finish notepad.exe from Android the call to *tasklist* shows running processes with their ID and *taskkill* then can terminate the process - just like the task manager on the local Windows machine itself.

2.3   REALTERM (WINDOWS)

This book uses the terminal software RealTerm as a software connection block. As many other applications of that kind it supports both the old RS232 serial interface COM and the TCP/IP interface on a Windows machine. Even the old Windows Hyper Terminal can handle both connections, as can be seen elsewhere in this book, but the highlight of RealTerm is the ability to redirect data to different serial channels. In addition, the free program has other features such as programmability, but the main aspect in this book is the crosswise connection and the telnet server.

RealTerm is like a software ESP8266 running on Windows, as both allow the connection of RX/TX and TCP/IP. Relevant examples can be found in many sections in this book. On the other hand, it can be used as a kind of *NETcat* with an easy to use interface. As Bluetooth uses a COM in Windows, Realterm can handle and log Bluetooth data.

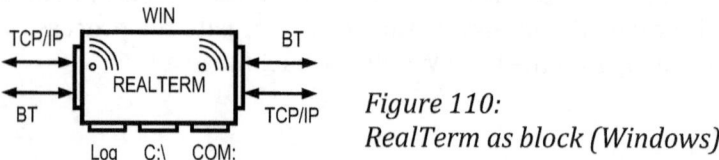

Figure 110:
RealTerm as block (Windows)

In the upper part of the screen the terminal window is found. That is where inputs and outputs are monitored. There are ten tabs for the various functions: *Display Port, Capture, Pins, Send, Echo Port, I2C, I2C I2C 2 Misc* and *Misc.* From the tab I²C special features of the program can be used which will not be further discussed or used here.

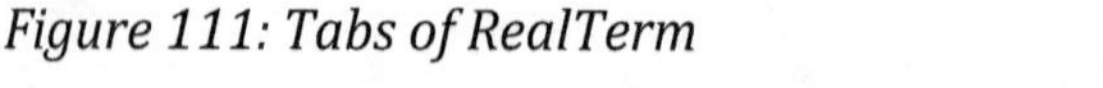

Display | Port | Capture | Pins   Send | Echo Port | I2C | I2C-2 |

Figure 111: Tabs of RealTerm

Display allows various settings for the display of data. Character set, duplex, hexadecimal, binary, ASCII just to mention a few. *Port* establishes the main link: the classic RS232 serial connection, a telnet-server and clients of the localhost (127.0.0.1). All the usual parameters of the interface can be set here. *Open* connects the selected interface. Incoming data can be stored in a text file by *Capture* continuously; there are various parameters available for this option.

The terminal window will no longer be updated, if data is caprured. The individual *Pins* of a connected/open RS232 port (RTS/DTR/TXD) can be toggled. *Send* allows individual bytes or characters or entire files to be transferred via an open connection. The adjustable delay between characters is very useful if using slow hardware. Other parameters allow individual transfers.

However, the highlight of RealTerm is the tab entitled *ECHO PORT*. Incoming data set under *Port* connection is conducted to a different port or even telnet-served. The setting here is identical, so e.g. COM1 with COM5 can be connected. But the connection of COM1 with a telnet server in order to extend further the serial data via network protocol is possible. An other application is TCP/IP data of an executable that is using the localhost (127.0.0.1) only (Sorcerer) to pass data to a device in the common network. In Chapter 3 of this book, this method is in use where the software decoder Sorcerer decodes its data obtained through the sound card transferring via TCP/IP port 55 on localhost 127.0.0.1. *RealTerm* picks that data at the incoming port, displays these bytes or logs them to a file and sends this information via the echo port to a phone or tablet, so this is routing through the local network by e.g. 192.168.178.23 (router-example) to do more things with the data on other machines.

2.4   JAVASCRIPT AND BT93

The JavaScript library *bt93.js* is a port of an old Turbo Pascal unit, which was in use when Windows was not very common on a PC and many users did not even use a mouse in wordprocessing. In 2012, with appearance of HTML5 and its canvas, it was published at:

http://www.hjberndt.de/soft/canbt93.html

The library is located in this path for download or access. This book uses this library in the section *ESP8266BASIC* to create simple charts in a browser. The few routines and constants are briefly explained at this point. Here's an example:

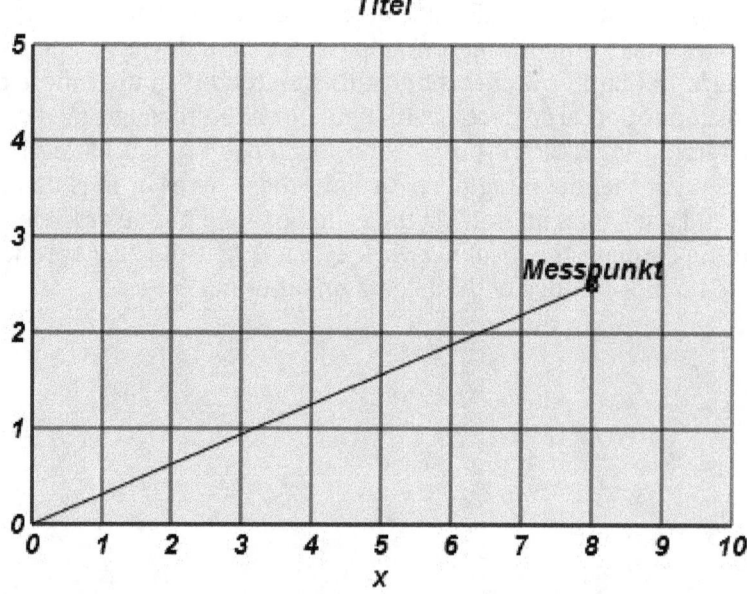

Figure 112: *Simple graph showing measurement point and line*

The HTML source code looks like this:

```
<!DOCTYPE html><html><body><br>
<canvas id="myCanvas" width="480" height="320" >
</canvas>
<script src="bt93.js" type="text/javascript">
</script>
```

```
<script type="text/javascript">

Grafik(AN);
farbe=WEISS;cls();
farbe=HELLGRAU;    hintergrund();
farbe=SCHWARZ;     xachse="x";yachse="";
Diagramm(0,10,0,0,5,1);
marktyp=KREUZ|KREIS;
DiaPunkt(8.0,2.5);
DiaLinie(0,0,8,2.5);
DiaText(8,2.5,"Messpunkt");

</script>
</body>
</html>
```

And here a brief reference of the functions (mostly German names), variables and constants of this simple library.

Grafik *(AN/AUS)*

Turns to graphics mode (AN) or off and initializes the necessary variables. AUS is not used.

Diagramm *(xa, xe, xs, ya, ye, ys)*

A chart, a grid and two axes are drawn. The parameters are the same for x and y and are the start, end, and step. If step is 0, then a 10 pitch is used. Using a logarithmic axis the third parameter is ignored. The strings *xachse*, *yachse* and *titel* may contain appropriate identifiers. The variable *diagrammtyp* (chart type) is pre-set to XLINYLIN for a double-linear diagram. With the constants XLINYLIN, XLINYLOG, XLOGYLIN and XLOGYLOG this global variable can be assigned to chart types before *Diagramm* is called.

DiaPunkt *(x, y)*

A point is plotted in chart units using the *marktyp* and the *markgroesse* if *Diagramm* has been called. The *markgroesse* (default 3) is measured in pixels. The *marktyp* can be a combination of KREUZ (CROSS), RAUTE (DIAMOND), KASTEN (BOX), DELTA, PUNKT (DOT) and KREIS (CIRCLE). These constants can be added or ored. The point appears in the color set to the *farbe* variable.

DiaLinie (x1, y1, x2, y2)

A line is drawn in chart units using the last color assigned to the variable *farbe* if *Diagramm* was called before.

DiaText (x, y, s)

The string *s* as a constant or variable is drawn at chart point x/y. For colour, the same applies as for point and line.

cls()

Clears the entire canvas or paints the canvas with the last colour.

hintergrund()

Clears the chart area (background) using the last colour.

printxy(x, y, s)

As *DiaText*, but in Canvas units (pixels).

draw(x1, y1, x2, y2, color)

Like *DiaLinie* but using canvas units and without the variable *farbe*.

plot(x, y, color)

Puts a pixel using the colour of *color* at canvas position x/y.

Global variables:

diagrammtyp, marktyp, markgroesse,

xachse, yachse, titel, farbe,

xwork, ywork, wwork, hwork

The last four variables hold the chart area in pixels units. Another example with logarithmic scale just for fun:

```
<script src="bt93.js" type="text/javascript"></script>
```

```
<script type="text/javascript">

Grafik(AN);
farbe=WEISS;cls();
farbe=GRAU;
xachse="x";yachse="";
diagrammtyp=XLOGYLOG;
Diagramm(1,10000,0,1,100,0);
marktyp=KREIS+KREUZ+RAUTE;
farbe=GRUEN;markgroesse=50
DiaPunkt(200,50);
DiaLinie(10,65,1000,65);
DiaText(100,70,"Messpunkt");
printxy(xwork,ywork-20,"log y");
</script>
```

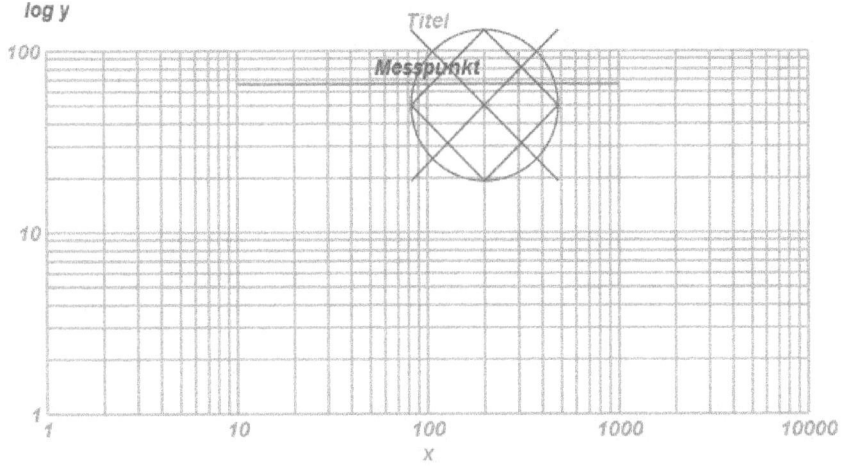

Figure 113: Logarthmic charting using BT93- JavaScript

2.5 RFO-BASIC (ANDROID)

A Google for "*rfo*-Basic" will get you to:

http://rfo-basic.com/

The bottom of the page says: "*This project is a labor of love by the curator of The Dr. Richard Feynman Observatory and author of Atari BASIC and Apple DOS 3.1: Mr Paul LAUGHTON. It is free to all, now, and forever. The underlying source code is available under the terms of the GNU General Public License.*"

The app is available in the PlayStore and runs on Android devices. Meanwhile, there are easy ways to generate an APP/APK from BAS files. The NetCompactServer for NetCompact client in this book was created like that. This BASIC was the main reason for writing the first eBook [1] in this series. In there, the possibilities of an Android-Smartphone have been shown and the simplicity of hardware access using this BASIC. In [2], the main focus was Bluetooth, data transmission and control, while *rfo*-Basic solved the needed connections in few lines of code.

This book focuses mainly on wireless and TCP/IP and again *rfo*-Basic shows its strength in the short and effective source code on the Android platform. All examples in this book are modifications of the examples provided by Paul Laughton. As a further aid the current documentation is recommended for download as a PDF file or to read the online web version.

2.6   NETCOMPACT (WINDOWS)

This section, as part of the work on this book, is published in essential parts as a supplement to the blue eBook at:

http://www.hjberndt.de/soft/netcompact.html

At the end of the last century data acquisition often was done via a serial RS232-interface. The Compact software was created in different versions and for various platforms. In this book a variant can be found in the chapter about the Digispark where Compact-Definition is used to perform measurements and control by the tiny board on a Windows PC.

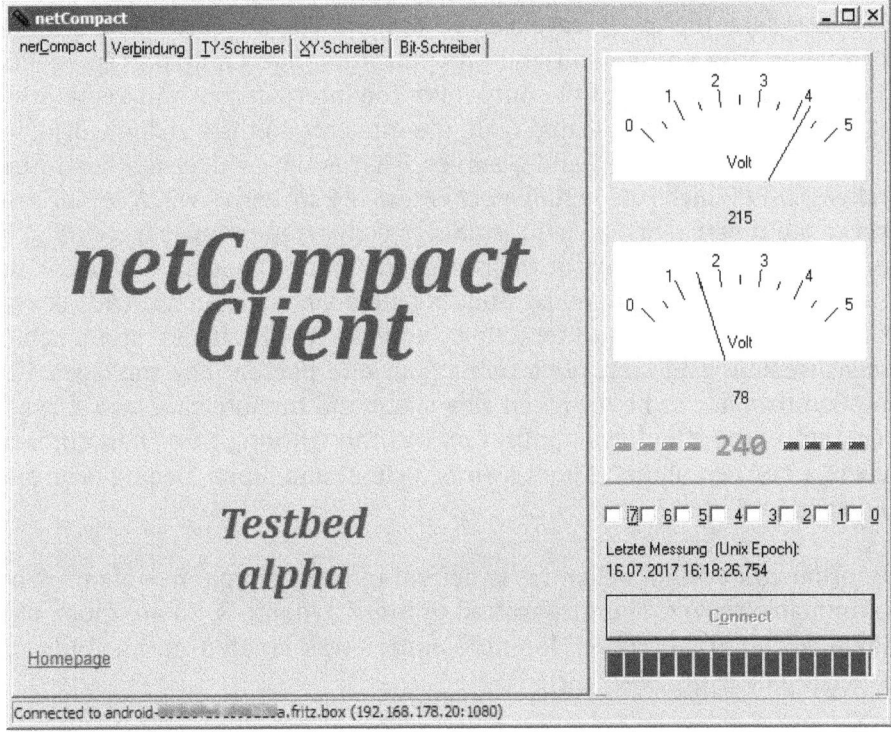

Figure 114: Test application for TCP/IP measurements

Due to the arrival of the *ESP8266* and the ability to use TCP/IP connections in measurement and control, this kind of device connection has become available to all. So the old serial RS232 connection will probably be replaced by Wi-Fi and sockets. The transmission method brings up new opportunities but some issues or problems too, to be practically analysed. To accomplish this, the TCP/IP version of Compact emerged. The interplay between the new Compact variant *netCompact* with appropriate hardware such as the *ESP8266* is a possible object of investigation with this first version. As a first step a server was built on an Android smartphone or tablet to watch the new connection and its special behaviour within netCompact.

2.6.1 CLIENT OR SERVER

The RS232 connection has two participants. In *CompactDefinition* and its predecessors, the tasks were clearly defined. The interface provides the computer data on demand and an application like Compact represents this data. Any control of the outputs of the interface are initiated by the computer and communicated from the interface to the outside world. In Sockets there is a client and a server. That is why a decision has to be taken, i.e. in which role *netCompact* enters the socket world. A server can serve multiple clients; a client is always connected to one server only. If an interface as a server of measurement data is a server in terms of sockets, it can be accessed by multiple clients to share data. Thus could the interface perform interaction between their clients and enable measurement and control to more than one person. For this scenario netCompact has to be designed as a client and the interface as a server. Therefore, the Windows application has the working title "netCompact client - Testbed alpha". Client as it is a client and alpha, beeing far from finished.

In order to develop a client, it is helpful to use a simple flexible or programmable server. The turnaround times of Arduino & Co are much too long, so initially a server for netCompact was created on an Android smartphone using *rfo*-Basic!

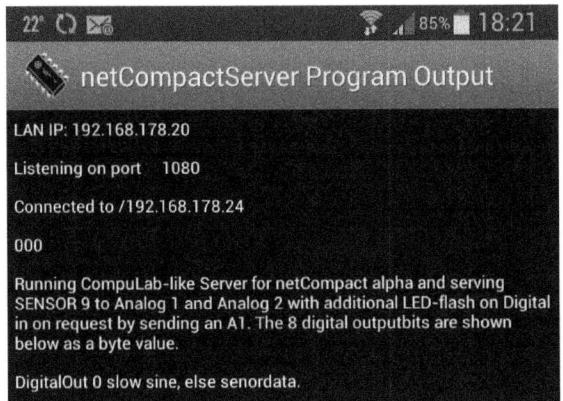

Figure 115: Android Smartphone/Tablet as a Serverfor netCompact

A first server on an Android smartphone provides sensor data to net-Compact. A compiled version of the *rfo*-Basic server (source below) with its simple interface is available as an Android app which serves its measurement data at (1080 by default) an adjustable port in a later on explained data format on the author's homepage. After entering the port number, the app displays its own IP and waits for a client. At connection a greeting is sent and then the loop is entered to respond according the incoming requests. The smartphone provides either two very slow changing amplitude values for checking the time behaviour or values of the position sensor in two dimensions. Switching is done for simplicity via the digital outputs of the simulated device. So if one of the 8 output bits is turned on, sensor data is supplied, else amplitudes.

If a server is running using a known IP, netCompact is able to connect and automatically send the measurement request "A1". A single sending of characters is done by pushing the send button, also line feed and carriage return can be set accordingly. The lower window outputs the data which may be stored in a log file. There is also an INI file that holds some settings. Registry entries are intentionally omitted. The log data is appended so very huge files may be written to disk. This log file can be destroyed by hitting "Delete".

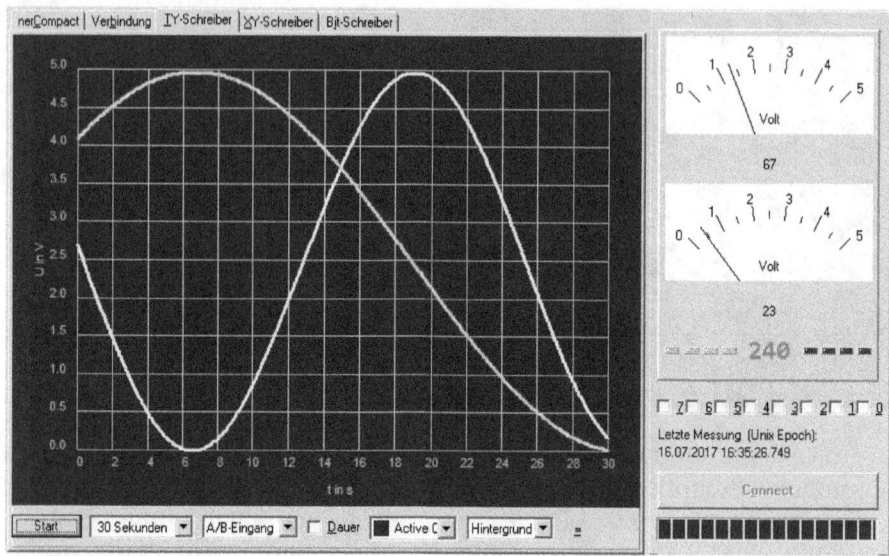

Figure 116: netCompact displaying Smartphone-values

The "Connect"-button tries to connect and if successful the progress bar will show 50%. Hitting the first measurement data the bar is displayed in full width and the received data will appear in the control window.

2.6.2 DATA FORMAT

The ASCII data format in this version is like this:

][1476694592256,0025,0040,0015

There are two brackets as reserved bytes, followed by the Unix timestamp using 13 characters, followed by comma-separated metrics, each with 4 characters, where currently only the last three digits are evaluated as an 8-bit value. To set both instruments to 2.5 Volts and to lighten only the lower four LEDs, a netCompact client has to send the following string via TCP/IP with a trailing newline (#10) :

][0000000000000,0127,0127,0015

13 zeros as a time stamp - that is January 1, 1970 - the beginning of the Unix epoch. Thus now it should be possible to do own transducers, for

example communicate using an *ESP8266*. Expanding the window at its lower border, the testbed gets visible. The time interval can be changed manually. Information is visible of measuring points and measuring point. The operation of the slider is in the sole responsibility of the user, especially towards smaller intervals. Both the apps are available for download to start own experiments at:

http://www.hjberndt.de/soft/netcompact.html

On first start-up of *netCompact* Windows might display a warning, but this may be ignored depending on the version of Windows. On an Android device the checkbox has to be checked in settings 'Security' at 'Unknown sources' to allow an APK-install even though it did not download from the Play Store.

Using the Smartphone Server, a circle can be drawn by phone wireless. One of the checkboxes has to be checked at the digital outputs for readout of the position sensor 9 of the used Android device being transferred. If the device is located with the display flat on a table, both hands of the instruments should point to the middle. Now holding the unit vertically in front of the face, the upper meter shows full scale and the lower display half rash. According the measuring point is located at "3 o'clock", or "0 degrees". By starting the recorder and turning the Android device slowly clockwise or vice versa, a circular figure is created by the superposition of the two movements.

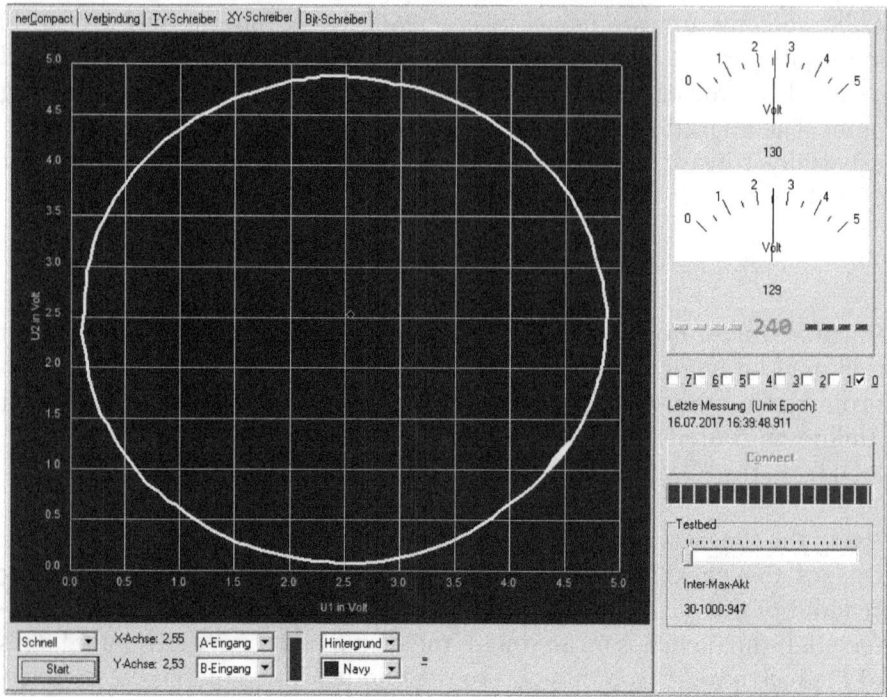

Figure 117: netCompact and sensor-9 data from an Android smartphone

2.6.3 RFO-BASIC!-SERVER

The easy to use *rfo-BASIC!Quick-APK* tool *rfo*-BASIC is able to compile a Basiclisting to an APK, called APP for the Android platform. The downloadable version of the server is the compiled listing shown below. The first eBook in this sequence was inspired by the simplicity of *rfo*-BASIC and accessing the sensors of the Android smartphones [1]. This server is originated from the demo file *ServerDemo*, coming with the *rfo*-Basic package.

```
out$="000"
SENSORS.OPEN 9
PAUSE 200

FN.DEF ain$(out$)
  t=CLOCK()/1000
  IF out$="000" THEN % SCHWINGUNGEN
   a=127+127*SIN(2*PI()*0.04000*t)
   b=127+127*SIN(2*PI()*0.02050*t)
```

```
ELSE
 SENSORS.READ 9,b,a,c
 b=127+b*12.5:a=127+a*12.5
 b=MAX(0,MIN(b,255))
 a=MAX(0,MIN(a,255))
ENDIF
IF MOD(t,2)<1 THEN c=15 ELSE c=255-15
A$="][" +FORMAT$("%%%%%%%%%%%%%%",TIME())+","+FORMAT$("%%%%",a)
+","+FORMAT$("%%%%",b) +","+FORMAT$("%%%%",c)+CHR$(13)
 a$=REPLACE$(a$," ","")
 FN.RTN a$
FN.END

FN.DEF send(l$)
 SOCKET.SERVER.WRITE.LINE l$+CHR$(13)
 FN.RTN 1
FN.END


INPUT "Enter the port number", port, 1080
SOCKET.SERVER.CREATE port

start:
!*********** From Server Demo  ********
SOCKET.MYIP ip$: PRINT "LAN IP: " + ip$
p$=FORMAT$("######",port)
PRINT "Listening on port ";p$
SOCKET.SERVER.CONNECT 0
! WAIT FOR CLIENT
DO
 SOCKET.SERVER.STATUS st
 PAUSE 100
UNTIL st = 3
! Client connected
ok=0
SOCKET.SERVER.CLIENT.IP ip$
PRINT "Connected to ";ip$%   CLIENT CONNECT
CALL Send ("Hi, I'm Eliza, tell me about your problem within 32 se-
conds.")
! GREETINGS FROM SERVER
PAUSE 2000

!SERVER LOOP
DO
 maxclock = CLOCK() + 30000
 DO
  SOCKET.SERVER.READ.READY flag
  IF CLOCK() > maxclock THEN
   PRINT "Timeout after 30 seconds."
   CALL send ("Timeout, thus not listening anymore and closing down.")
   ONERROR:
   SOCKET.SERVER.DISCONNECT
   PRINT "Disconnected from client"
   GOTO start
  ENDIF
  IF flag =0 THEN PAUSE 20 % CPU/BATTERY
 UNTIL flag
 SOCKET.SERVER.READ.LINE line$
```

```
SW.BEGIN LEFT$(line$,2)
 SW.CASE "A1": CALL send(ain$(out$)): SW.BREAK
 SW.CASE "D1": CALL send(ain$("D1")): SW.BREAK
 SW.CASE "O1": Out$=RIGHT$(line$,3):  SW.BREAK
 SW.DEFAULT :  CALL send("funky"): PRINT "FUNKY "+line$: SW.BREAK
SW.END
IF old$<>out$ THEN PRINT out$
old$=out$
PAUSE 2
IF ok =0 THEN PRINT "Running CompuLab-like Server for netCompact alpha
and serving SENSOR 9 to Analog 1 and Analog 2 with additional LED-flash
on Digital in on request by sending an A1. The 8 digital outputbits are
shown below as a byte value.":print "DigitalOut 0 slow sine, else
senordata."
 ok=1
UNTIL true
END
```

3 COMPOSITION/INTERPLAY/INTERACTION

This section shows some arbitrary but concrete examples of the interaction of the various components and methods described previously. It is shown how a smartphone or tablet is enabled to communicate with the Digispark wireless by TCP/IP. Another example is a talking clock on the phone, receiving the information by a shortwave radio in digimodes - a type of ATC time announcement. As a kind of Chinese whispers with voice recognition and voice output on Android and Windows is shown in other script examples. ESP8266Basic is used for measurement and charting data even of more than one analogue source. But first, the smallest Arduino in the world: Digispark.

3.1 DIGISPARK - TCP/IP (WLAN)

The Digispark connects natively via USB and/or RX/TX (TTLSerial). One of these paths must be taken to talk via Wi-Fi, if not reinventing everything via I/O itself at the very lowest level. Since the Board-USB is very slow, the connection via SoftSerial is best. As explained in the first part of this book, there are several usable hardware components.

- HC06 Serial/Bluetooth >   (Win) > Wi-Fi

- FTDI Serial/USB/Serial >   (Win) > Wi-Fi

- ESP8266 Serial/W-iFi

- HC06 Serial/Bluetooth >   (Android) > Wi-Fi

Depending on the selected method, appropriate software and/or hardware components are required on the other end of the communication.

3.1.1 DIGISPARK: HC06 SERIAL/BLUETOOTH > (WIN) > WIFI

This variant comes without the use of an *ESP8266*. The construction uses the known *HC06* module to let the Digispark transmit serial data via Bluetooth. A smartphone could now read that data - as above via Bluetooth -, but here the data is to be spread wireless and perhaps by internet into the whole big world.

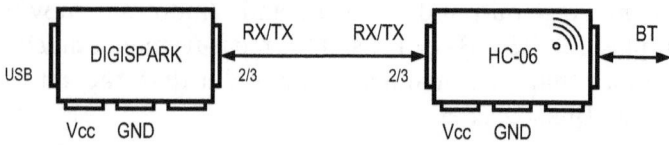

At receiver side, a Windows tablet is used to achieve the conversion of Bluetooth to TCP/IP using free software. The tablet can then be connected locally using a smartphone or an iPod by a router or even to the Internet. As schematic blocks might look like this:

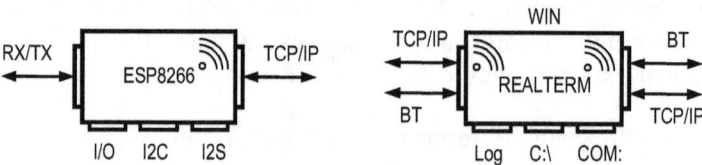

In some parts the *RealTerm* software does the same job as the hardware module *ESP8266*, since both are capable of redirecting serial data to TCP/IP. This property is to be used now. As noted in [1], Windows can communicate via a standard COM interface to Bluetooth. Thus *RealTerm* connects via e.g. COM5 to the Bluetooth transmitter *HC06* connected to the Digispark. The echo function in *RealTerm* forwards the data to e.g. Port 23 (telnet). The smartphone uses a TCP/IP client and connects to the Windows tablet through a router on port 23. The result is a TCP/IP connection using the little Digispark.

To test this configuration a Digispark sketch as listed below is sending "Hello Digispark" as well as the elapsed time in ms once a second, by using *SoftSerial*.

```
#include <SoftSerial.h>
#include <TinyPinChange.h>
```

```
#define  RX    2 //BluetoothSerial
#define  TX    3 //BluetoothSerial

SoftSerial mySerial(RX, TX);
#define Serial mySerial

void setup()
{ Serial.begin(9600);
}

void loop()
{ Serial.print("Hallo Digispark\t");
  Serial.println(millis());
  delay(1000);
}
```

If RX/TX (Pin 2/3) is tied to TX/RX (crosswise) to the HC06 module the transmitter is ready to use.

The Windows tablet is coupled via Bluetooth to a *HC06* module, but not connected yet. The corresponding Bluetooth/COM interface (e.g. COM4) is opened in the terminal program, whereby the connection procedure takes about two seconds. The LED flashing of the *HC06* module changes to continuous illumination as an intermediate result, the release of Digispark in the terminal should be legible. Now, in the tab *Echo Port* Port 23 (server: telnet) is selected and initiated with *Echo On* and *Change* the forwarding. Corresponding green LED replicas indicate a successful connection. Now the receiver is ready to use too.

On the smartphone, or any other TCP/IP-enabled device, a TCP/IP client is used. (For lack of hardware that works fine on the win-system using HyperTerminal to test). Here are the results on a Galaxy Note on Android.

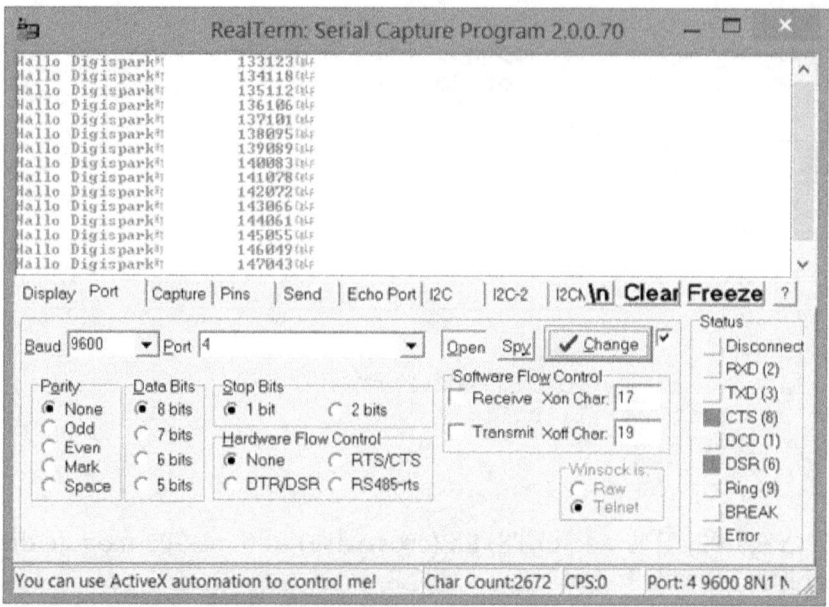

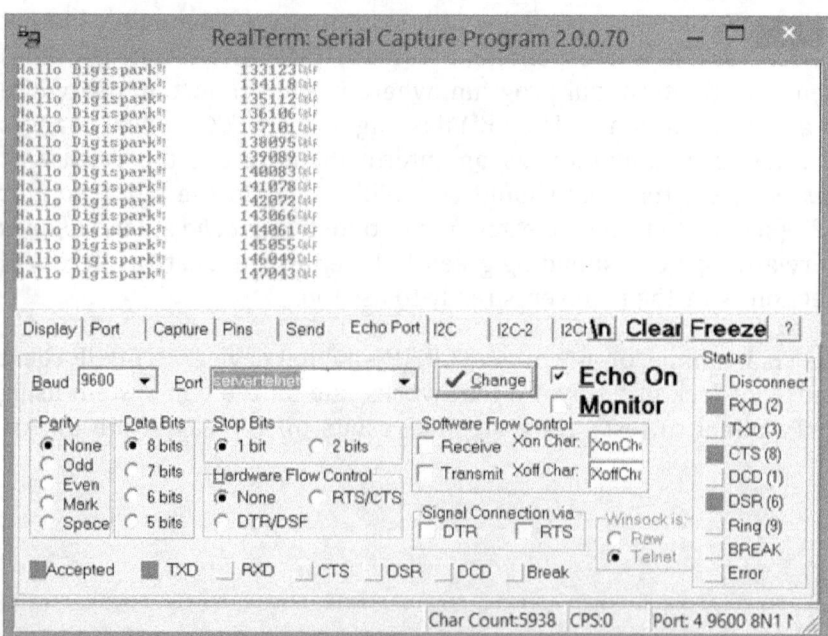

Figure 118: Bluetooth adapter HC06 shares its data on COM4: 9600 baud with and RealTerm manages the Bluetooth data echoed to Port 23

Adaptername:	Wi-Fi
SSID:	E5830-c818
Verbindungstyp:	802.11g
IPv4-Adresse:	192.168.1.101

Figure 119: The Windows tablet and Wi-Fi information in the Task Manager

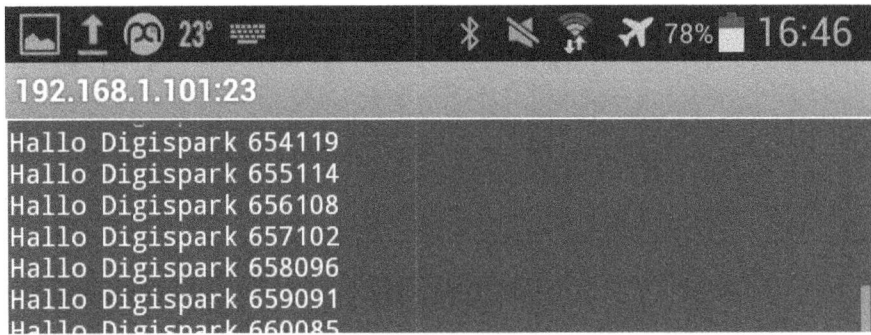

Figure 120: Android TCP/IP client is connected to the Digispark to a Windows tablet

Using RealTerm and NetCat

Using *NetCat* may seem a bit cumbersome, but aim of this book is to show possible solutions to individual problems in a modified approach.

Again *RealTerm* gets data via *HC06* from the Digispark using COM5. This time no echo function is in use, so other terminal applications can be used, if not supporting this *RealTerm* feature. However, using this solution data has to be available in a log file, so the app has to save the incoming data as text, which is usually given. In *RealTerm* the *Capurte.txt* file can monitor incoming data or save it. *NetCat* reads this log file and sends its content via TCP/IP to the smartphone. In detail, on Windows it could look like this:

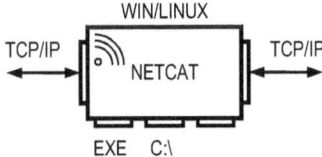

The incoming data is in the log file *C:\temp\CAPTURE.TXT*. The tablet is logged by the router with e.g. the local IP 192.168.1.101. On the Windows Tablet, the editor (Notepad.exe) creates a text file with the following content: *C:\temp\nc\nc -L -p 33 -e cmd.exe*. The file is stored and renamed as *digi.bat* (icon changes). Before this file is executed it has to be true that *nc.exe* lives in the directory (folder) *C:\temp\nc*.

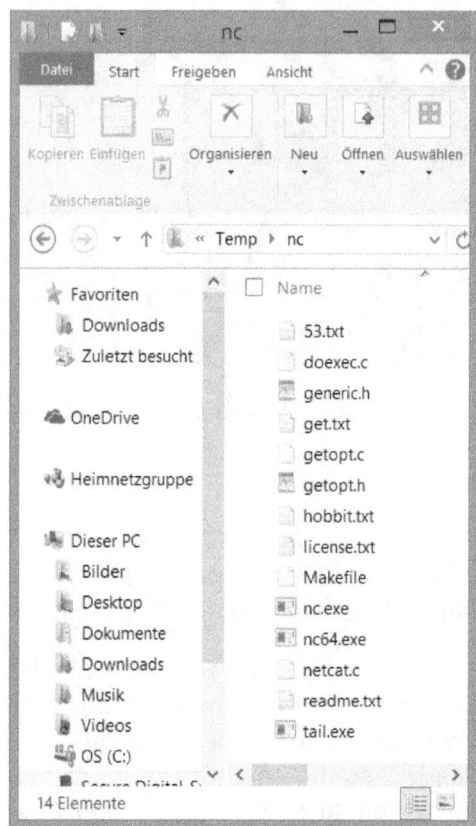

Figure 121: The old C:\Temp directory is well suited for experiments. NetCat and tail in the subdirectory nc. NetCat has to be logged in to the firewall (Win8.1, Administrator)

The BAT file then calls *NetCat* (nc.exe) in such a way that the tablets will listen on TCP/IP port 33 for someone to connect. Once this occurs, the command processor of Windows is run and all in- and output is redirected to port 33. This is the way an Android smartphone can take over the Windows tablet.

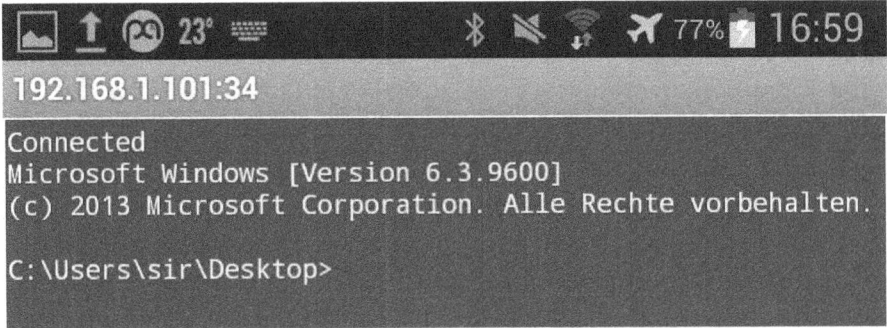

Figure 122: Windows controlled from a smartphone using NetCat

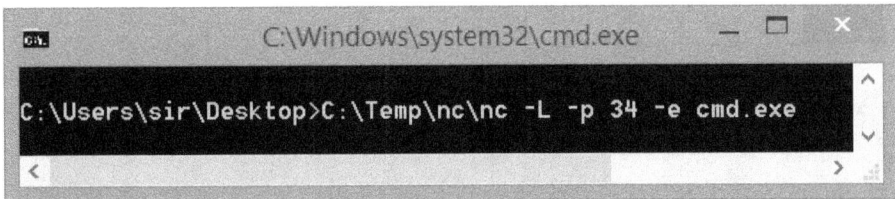

Figure 123: Calling NetCat from command line

This is where the file *CAPTURE.TXT* is to be sent. As there are often such changing files in the digital world, there is a simple tool to manage such files continuously. The Windows Resource Kit SDK Windows 2003 - still available - comes with *tail.exe*, which transfers the last 10 added lines of a changing file. If this file is in the same directory as NetCat, it can be called accordingly. (Alternatively, VBScript can be used to code such a function) On the smartphone, the client is started at the Windows prompt using the following call:

```
C:\Temp\nc>tail -f c:\temp\capture.txt
```

and the data flow from Digispark should arrive in a 10 line interval on the smartphone.

Summary: Digispark Serial> Serial Bluetooth> Bluetooth COM> (Real-Term)> COM file> (NetCat/Cmd/Tail)> File TCP/IP

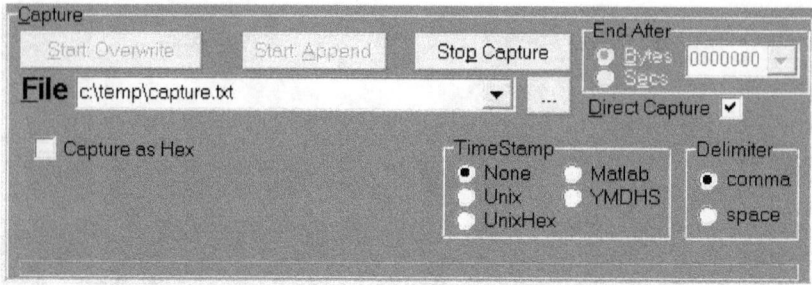

Figure 124: RealTerm stores in its log file CAPTURE.TXT

Since *tail*.exe is in the same directory as *nc*.exe and this directory is not added to the Windows PATH variable, the entire path needs to be entered, or *cd..* to that folder like in the old DOS-days to get there.

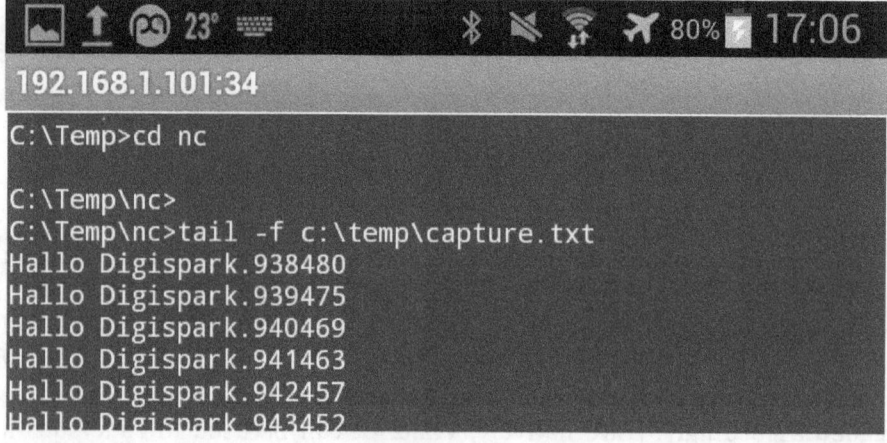

Figure 125: Call and result of tail.exe on the phone

3.1.2 DIGISPARK: SERIAL TO FTDI/COM > COM/TCP/IP (WI-FI)

In order to connect to *RealTerm* using the *FTDI-Adapter* by wire, the *HC06* module has to be replaced by this hardware.

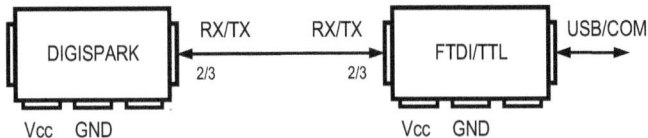

The sketch and the wiring on and around Digispark can remain the same. Except for the replacement of hardware, only the number of the COM port which is set in *RealTerm* tab *Port*, varies substantially. Now both methods described above should work unchanged.

3.1.3 DIGISPARK: SERIAL TO ESP8266 > TCP/IP (WI-FI)

The smallest and cheapest solution comes without a Windows helper. The module *ESP8266* was constructed to redirect serial data to TCP/IP or Wi-Fi. Whether it makes sense still using a Digispark having the capacity of an ESP, can be discussed or not - only the opportunity is to be shown.

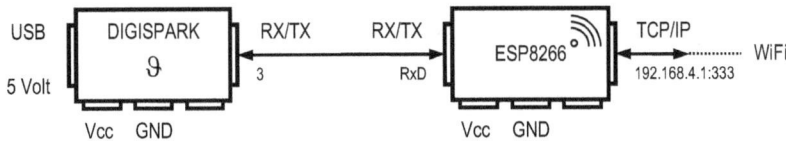

Figure 126: Transfer of the inner Digispark temperature alternatively via Wi-Fi

The Digispark sketch and the connection remain unchanged. RX/TX however now is connected to TX/RX of the ESP. A SerialToTelnet sketch from examples does the job. The Windows/RealTerm variant is in fact replaced by this programmable device.

The next sketch for the *ESP8266/12* is a modification of the example of the ESPWiFi-library (Core) and connects the serial port of an ESP and the TCP/IP (Wi-Fi) interface. It is coded using an Arduino IDE.

In ESP WiFi-AP-mode no other infrastructure is needed. The smartphone connects to a "hot spot" with the name *ESPap* and password-free access. A TCP/IP client can now connect via IP using port 333 and read the messages of Digispark.

These lines connect Serial and Wi-Fi using an *ESP8266*:

```
#include <ESP8266WiFi.h>
#define PORT 333
#define MAX_SRV_CLIENTS 3

WiFiServer server(PORT);
WiFiClient serverClients[MAX_SRV_CLIENTS];

#define Serial1 Serial

void setup()
{WiFi.disconnect();
 delay(100);
 pinMode(13, OUTPUT);Serial1.begin(9600);// pin 2 tx
 WiFi.softAP("ESPap", "");
 Serial.begin(9600);
 server.begin();
 server.setNoDelay(true);
 Serial1.print("Ready! Use 'telnet ");
 Serial1.print(WiFi.softAPIP());
 Serial1.println(" 333' to connect");
}

void loop()
{uint8_t i;
 if (server.hasClient())
 {for(i = 0; i < MAX_SRV_CLIENTS; i++)
  {if (!serverClients[i] || !serverCli-
ents[i].connected())
   {if(serverClients[i])serverClients[i].stop();
    serverClients[i] = server.available();
    Serial1.print("New client: "); Serial1.print(i);
    continue;
   }
  }
  WiFiClient serverClient = server.available();
  serverClient.stop();
 }
 // COPY TO SERIAL
```

```
for(i = 0; i < MAX_SRV_CLIENTS; i++)
{if (serverClients[i] && serverClients[i].connected())
 {if(serverClients[i].available())
  {while(serverClients[i].available())
   {char c = serverClients[i].read();
    Serial.write(c);
   }
  }
 }
}
// COPY FROM SERIAL
if(Serial.available())
{size_t len = Serial.available();
 uint8_t sbuf[len];
 Serial.readBytes(sbuf, len);
 for(i = 0; i < MAX_SRV_CLIENTS; i++)
 {if (serverClients[i] && serverClients[i].connected())
  {serverClients[i].write(sbuf, len);
   delay(1);
  }
 }
}
}
```

3.1.4 *DIGISPARK: SERIAL/BLUETOOTH > TCP/IP (WI-FI)*

The fourth variant reconnects Digispark to the *HC06* Bluetooth module, but the Windows tablet and *RealTerm* are replaced by an Android app. The software builder Next Prototypes, known from well programmed Bluetooth and TCP/IP clients in the Google Play Store, combines both transmission paths in his app *SerialTransfer*. In this context, it fulfils the function of *RealTerm*, on an Android platform. It acts like a kind of software *ESP8266* when it's about Serial to TCP/IP conversion. The serial data arrives on Android via Bluetooth and is redirected to a TCP/IP port ready to use. The Windows tablet is to be the recipient.

The Digispark still is sending its Bluetooth messages to the *HC06* module. The app *SerialTransfer* connects by Bluetooth. Now the Windows tablet can act as a receiver. In *RealTerm* a server is running for the Android phone that can now log on by the app *SerialTransfer* as a client to *RealTerm*. This is done by selecting telnet server in the *Port* tab.

Finally *SerialTransfer* can be connected to the tablet via TCP/IP (like 108.168.1.100:23).

As a result of this mapping it can be seen that somewhere something is stuck. Presumably the developer Next Prototype is using a faster Android as the elderly Galaxy Note 1 here in use. In a direct Bluetooth connection there are no errors using the tablet.

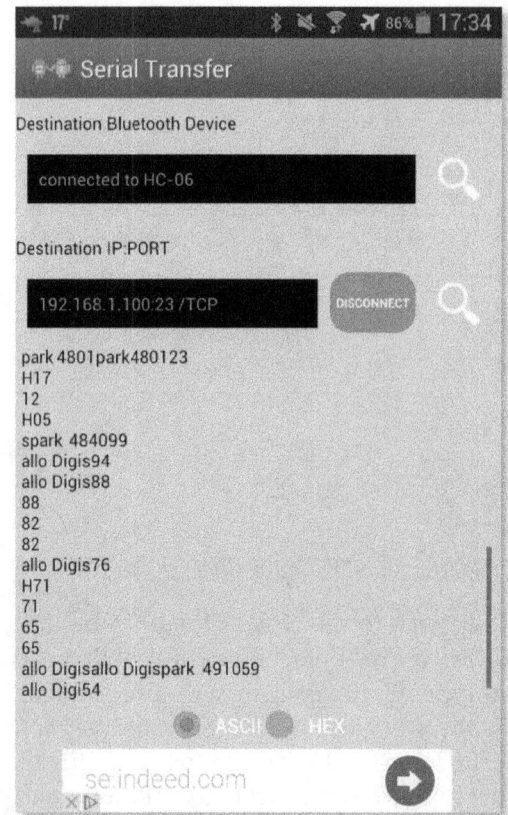

Figure 127: Connection Bluetooth to Wi-Fi on Android

3.2 ESPBASIC

In this section of this chapter further applications are presented by and with *ESPBasic*, which correspond to the context of this book.

3.2.1 ESPBASIC: SERIAL INTERFACE

Programming the serial interface is quick and convenient due to the event-oriented program control by *ESPBasic*. After things are initialized and configured, BASIC waits for the arrival of serial data, and then responds accordingly.

```
serialprintln "hi"
serialbranch [serialread]
wait

[serialread]
serialinput s$
serialprint "received: "
serialprintln s$
wprint s$
return
```

After a brief greeting on the serial channel, the branch target will be determined on serial data and serviced. On receiving the string *s$* is filled with the incoming data, which then are sent through the other connection with an acknowlege. If using the ESP module *Witty Cloud* connected by USB to Windows, or *HC06* via Bluetooth to another device with a serial terminal program (9600 Bd), a test can be run.

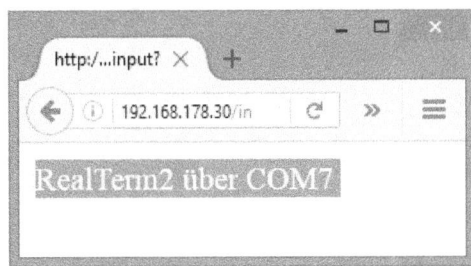

Figure 128: Message in browser and RealTerm

RealTerm in half-duplex mode is connected to the ESP module by COM7 and displays the data accordingly. Meanwhile BASIC and the Winsock reports status messages, as before the ensuing interaction was subjected to a test.

3.2.2 ESPBASIC: TCP/IP AND SERIAL INTERFACE

A BASIC property of the *ESP8266* is connecting the two serial data flows RX/RX and TCP/IP. Its RX/TX lines with TTL levels can take information from Wi-Fi and vice versa. The *ESP-core* for the Arduino IDE provides an example for telnet to realize this connection in C. The corresponding BASIC Listing is short and effective like this:

```
xtextbox T$
textbox S$
print "Hi. TCP/IP2Serial mit esp8266Basic"
if telnet.client.connect("192.168.178.26",23)>0 then
 telnet.client.write("Hallo. Hier spricht espBASIC!")
 Telnetbranch [clientread]
 serialbranch [serialread]
else
 print "Keine Verbindung."
 end
endif
wait

[clientread]
T$ = telnet.client.read.str()
serialprint T$
wait

[serialread]
serialinput S$
telnet.client.write(S$)
return

[ende]
telnet.client.stop()
end
```

Two textboxes display the TCP and serial mode data. The program uses event control and thus without constant checking the availability of data of the respective interface. It should be noted that as used herein Version 3.3 of serial Branch *serialread* ends with *return* - the socket Branch *clientread* however concluded using a *wait*. Whenever data applies, it is passed unchanged to the other interface. Finally, the server connection is established with the appropriate IP and closed again at the end by as much as two lines of code.

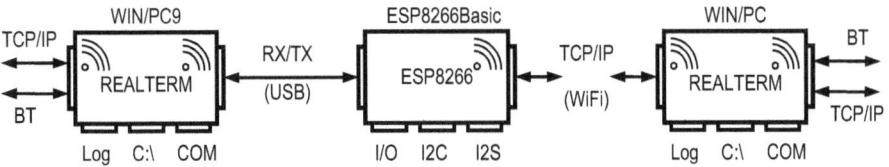

Figure 129:
Serial to TCP/IP using ESP8266 and RealTerm for testing

For testing this telnet to serial client, a telnet server is required. Besides Android apps and *rfo*-Basic on Android, a test can be performed on a Windows tablet using *RealTerm*. This program, described elsewhere, should be able to serve as a remote station for both communication channels. Two instances of RealTerm have to be started for this test. The first call establishes a connection to a telnet server with its own local IP or localhost. This server listens on port 23 and waits for the *ESP*. The second call to *RealTerm* serves as serial terminal on COM7 speaking via USB to the serial port of the ESP. The schematic block of this interaction could look like shown above.

If *ESPBasic* is running, the steps are as follows:

- Type/upload program into the ESP8266Basic editor of the browser
- Connect ESP 8266 by USB - (host) cable to the tablet (COM x)
- Run RealTerm1 and Telnet Server
- Run Realterm2 and connect with COMx
- Save Basic-Program (Save) and start (Run)

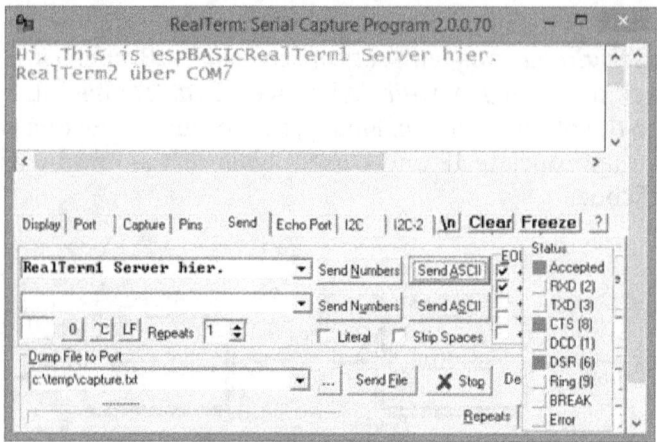

Figure 130: Server in RealTerm

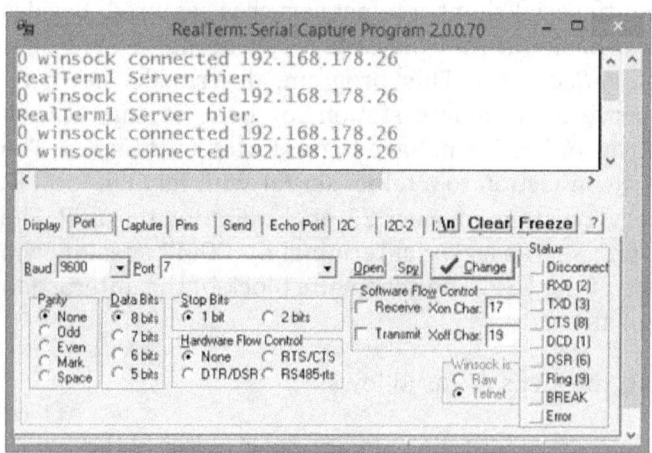

Figure 131: Serial connection via COM7:

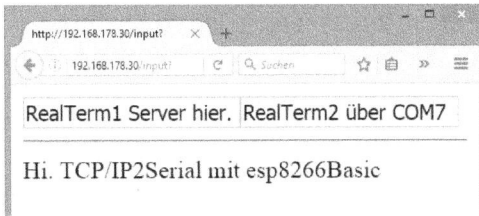

Figure 132: The mediator in a browser window

The *ESPBasic* program only acts as the mediator and behaves completely passive. The text displays are a copy of the data transmitted only and not for user input. Thus, the ESP module fulfils its fundamental task: TCP/IP Serial converter – using a few Basic lines.

3.2.3 ESPBASIC: *NETCOMPACTCLIENT*

The Windows application *netCompactClient*, as described in the software section of this book, originally designed for serial interfaces, should communicate with devices via TCP/IP or sockets. Here there is a corresponding minimalist version of this software written in BASIC, however, it only tracks the connection part. The display of measured values is achieved more below.

To connect and disconnect incoming data some more lines are needed, but still far less compared to the Delphi written genuine Windows application.

The program has four parts. At the start variables are initialized and components are placed on the browser window. First textboxes are placed for displaying messages and incoming data. Next the time stamp, which is converted from the transmitted data using BASIC routines in a readable date, follows. Below there are two original analogue data displays and a digital input. The interval of sending the measurement request "A1" is set by a drop box. At the end of initializing the connection to the server, in this test the Android app *netCompactServer*, is called by the router distributed local IP. Finally there are the determination of the jump destination for incoming data, and the start of the timer.

The second part *mach* will run at each timer event. The timer job is to send an "A1" to the measurement data providers, announcing a data inquiry. The connection is checked and optionally interrupted by a message or closed at this point.

The third part processes incoming data. After removal of line feed and carriage return, the decoding of the received string is done to appear as the time stamp and the three bytes in the text boxes.

The fourth part provides the digital outputs to "O10xxx" to switch as an 8-bit value. Finally, there is the target for the Stop button.

esp8266Basic als netCompactClient

Zeigt die drei Byte-Werte. Beim Android-Server wird zwischen Sensor 9 und den Schwingungen umgeschaltet.

][1476568586001,0210	Sensors
15. Oct 2016 21:56:26	Zeitstempel
0210	
0240	
0240	
Intervall/ms: 1000 ⌄	

STOP

Figure 133: net-CompactClient in ESPBasic

```
dim m$
a1 = 256
b1 = 256
c1 = 256
m$ = "Connecting..."
x = 0
wprint "<hr noshade size=1><font size=5
face=Arial,Helvetica>esp8266Basic als netCom-
pactClient</font><hr noshade size=1>"
wprint "<font size=2 face=Arial,Helvetica>Zeigt die
drei Byte-Werte. Beim Android-Server wird zwischen
Sensor 9 und den Schwingungen
umgeschaltet.<br><br>"

textbox m$
button "Sensors",[dout]
wprint "<br>"
textbox stamp$
wprint " Zeitstempel<br>"
textbox a$
meter a1,0,255
wprint "<br>"
textbox b$
meter b1,0,255
wprint "<br>"
textbox c$
meter c1,0,255
bla = 1000
wprint "<br>Intervall/ms:    "
dropdown bla,"10,50,100,500,1000,5000,10000"
wprint "<hr noshade size=1>"
button "STOP",[ende]
if telnet.client.connect("192.168.178.23",1080)>0
then
telnet.client.write("Hi. This is espBASIC")
telnet.client.write(chr(10))
telnet.client.write("O10000" & chr(10))
Telnetbranch [clientread]
else
 print "Keine Verbindung."
 end
```

```
endif
timer 500,[mach]
wait

[mach]
t = bla
e$ = "Verbindung unterbrochen."
timer t,[mach]
if telnet.client.write("A1" & chr(10))=0 then
 m$ = e$
 wprint e$
 goto [ende]
endif
wait

[clientread]
r$ = telnet.client.read.str()
r$ = replace(r$,chr(10),"")
s$ = replace(r$,chr(13),"")
m$ = left(s$,20)
'wprint s$
if left(s$,2) = "][" then
 h$ = word(s$,1,",")
 u$ = mid(h$,3,10)
 a$ = word(s$,2,",")
 b$ = word(s$,3,",")
 c$ = word(s$,4,",")
 a1 = val(a$)
 b1 = val(b$)
 c1 = val(c$)
 d$ = unixtime(u$,"day") & ". " & unix-
time(u$,"month") & " " & unixtime(u$,"year")
 t$ = unixtime(u$,"hour") & ":" & unix-
time(u$,"min") & ":" & unixtime(u$,"sec")
 stamp$ = d$ & " " & t$
else m$ = s$
endif
wait

[dout]
```

```
if x then telnet.client.write("O10128") else
telnet.client.write("O10000")
telnet.client.write(chr(10))
x = not x
wait

[ende]
telnet.client.stop()
end
```

A test run using the Android server, recognizable by the time stamp, is shown above. In section *ESP8266 core* there is a C-listing for a netCompact server for the ESP module, which then delivers the analogue value of the on board LDR at one channel, while the other channel provides a slow oscillation, as in the Android server in *rfo*-Basic. Controlling the RGB LED can be done either.

3.2.4 ESPBASIC: ESP CHAT

If there are two ESP modules available running one of them the server sketch and the other module the Basic-Sketch from the section above, the same listing could be used for an interactive module communication.

It should be noted that both of them have the same default IP without additional router or hotspot ESP 192.168.4.1, if operating as standalone AP. Without changing the IP, however, work can get easy by using a mobile hotspot from a smartphone – aka tethering. On a Galaxy Note flight mode has to be disabled, mobile data is not required and may stay off. Depending on the system, the assigned IP can be found in the hotspot setting to take them into account in the listing.

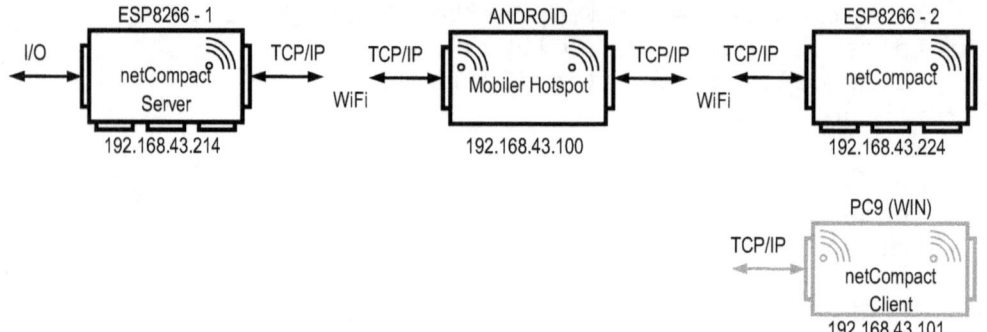

Figure 134: Two ESP in service of netCompact using a smartphone hotspot

In this test an old router serves as an intermediary using the two IP's xxx.xxx.xxx.31 for the client in BASIC and xxx.xxx.xxx.24 for the server in C. The corresponding line in the BASIC client contains the corresponding address:

```
if telnet.client.connect("192.168.178.24",1080)>0
then ...
```

After saving and running, the server logs in with its greeting, then remains hanging. The second run delivers the requested data through in the set 1000 ms interval.

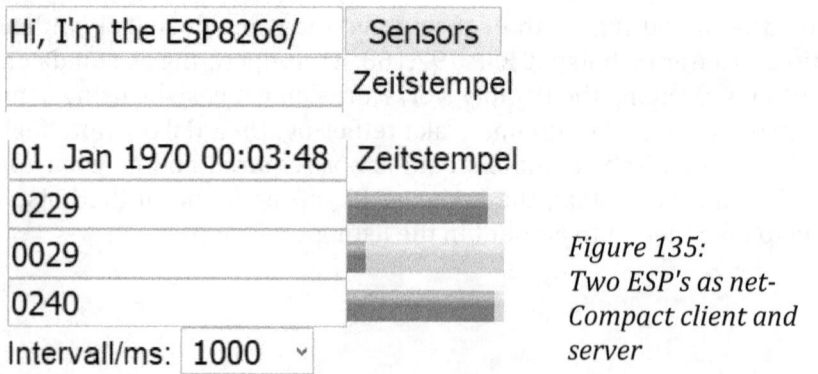

Figure 135: Two ESP's as net-Compact client and server

As no clock is running on the ESP, nor a time server is requested, the timing is the origin start of the UNIX time and then runs with attached milliseconds as a time stamp on port 1080 to the client.

3.2.5 ESPBASIC: HTML/JAVASCRIPT

In [1] it was decided not to reinvent charts or diagrams, but rather to show how HTML and *JavaScript* can be used for doing so. What was considered for *rfo*-Basic there, applies for reasons of space even more for *ESPBasic*. Since currently the "graphics" commands in version 3.55 are not quite doing what is expected here, this is the only way to present readings graphically. Stefan Thesen on his site shows since 2015 a web server by using Google's online libraries to get temperature and humidity displayed. In this section, however, the graphing is internet-independent or offline working. Connecting a DHT11 sensor to a *Witty Cloud* ESP and its on-board photo resistor, a display in the browser window by means of *ESPBasic* is possible like this:

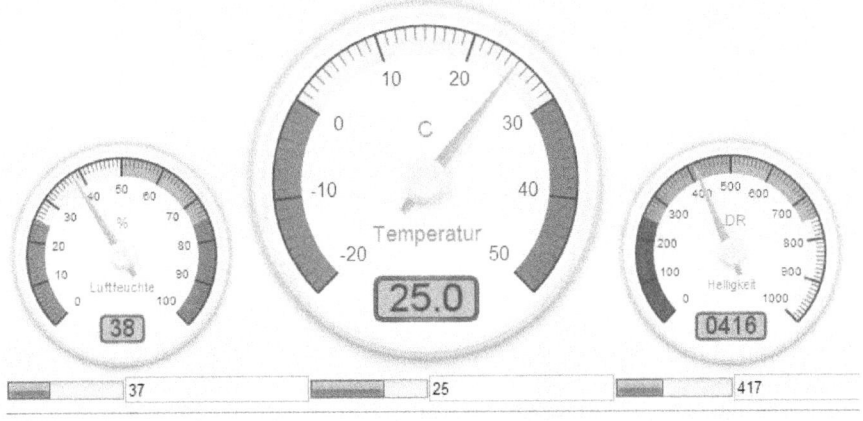

Figure 136: In the upper part there are three JavaScript Gauges, at the bottom the corresponding ESPBasic displays (meter and textbox); the meter element is a new available standard element in HTML5. All shown JavaScript elements are free software

3.2.6 ESPBASIC: JAVASCRIPT LIBRARIES

ESPBasic uses *JavaScript* itself and puts these packed gz-libraries into the file system of the *ESP8266*. Own libraries can be uploaded by the ESPBasic-*FILE MANAGER* using UPLOAD. They are stored in */uploads/* and are accessible like online *js* files. As an example, the clock from the website "HTML 5 and Canvas"

http://www.hjberndt.de/soft/canbt93.html

or concretely the URL

http://www.hjberndt.de/soft/BTUHR.html

is to be displayed locally. The source code of the HTML page (usually displayed using the right mouse button in a browser) looks like this:

```
<!DOCTYPE html>
<html>
<body bgcolor="black">
<canvas id="myCanvas" width="1280" height="720"
style="border:0px solid #c3c3c3;">
Canvas not found.
</canvas>

<script src="bt93.js" type="text/javascript"></script>
<script type="text/javascript">

function Ziffernblatt()
{var x,xx,i,j;
 x=0;
 for(i=1;i<=12;i++)
 {farbe= WEISS; DiaPunkt(Gsin(x),Gcos(x));
  xx=x;
  for(j=1;j<=4;j++)
  {xx=xx+360/60;
   farbe=GRAU;
   DiaLinie
(Gsin(xx),Gcos(xx),0.98*Gsin(xx),0.98*Gcos(xx));
  }
  x=x+360/12;
 }
}

function Zeiger(h,m,s,c)
```

```
{var a;
 switch(c)
 {
   case 'm': ctx.lineWidth=10;//SETLINESTYLE(0,0,3);
             a=(m/60*360);
             farbe=ORANGE;
             DiaLinie(0,0,0.95*Gsin(a),0.95*Gcos(a));
             a=m*360/12/60;
             a=a+(h/24*2*360);
             farbe=ORANGE;
             DiaLinie(0,0,0.75*Gsin(a),0.75*Gcos(a));
             ctx.lineWidth=3;//SETLINESTYLE(0,0,1);
             break;
   case "s": a=(s/60*360);
             farbe=ROT;
             DiaLinie(0,0,0.95*Gsin(a),0.95*Gcos(a));
             break;
 }
}

function Uhr()
{var h,m,s,now=new Date();
 Grafik(AN);markgroesse=4;ctx.lineCap="round";
 ctx.lineWidth=3;
     hwork=ctx.canvas.height-4*markgroesse; wwork=hwork;
     ywork=2*markgroesse;
     xwork=ctx.canvas.width / 2 - wwork / 2;
     farbe=SCHWARZ; Diagramm(-1,1,0,-1,1,0);
     farbe=WEISS;   Ziffernblatt();
     h=now.getHours();
     m=now.getMinutes();
     s=now.getSeconds();
     Zeiger(h,m,s,"m"); Zeiger(h,m,s,"s");
}

window.setInterval("Uhr()", 1000)
</script>

</body>
</html>
```

The HTML code has three parts: HTML Start - integration of the *JavaScript* library - user lines in *JavaScript* - HTML-end. Using the *ESPBasic* statement *Wprint*, or with Version 3 the shorter statement *html*, that kind of code is an easy task to construct. To prevent too many quotes in a line of *ESPBasic* the "|" character is accepted as the beginning and the end marking of the *Wprint/html* statement (no line breaks). Thus the output of the first three lines of the HTML page in *ESPBasic* looks like this:

```
html |<!DOCTYPE html>|
html |<html>|
html |<body bgcolor="black">|
```

The same applies to *JavaScript*, as shown below. The bolded line integrates the library *bt93.js* and has to be adapted for this environment.

```
html |<html><body bgcolor="whitesmoke"><br>|
html |<canvas id="myCanvas" width=480 height=320
style="border:0px solid #c3c3c3"></canvas>|
html |<script src="/file?file=bt93.js"
type="text/javascript"></script><script
type="text/javascript">|
html |function Zifferblatt()|
html |{var x,xx,i,j;|
html |  x=0;|
html |  for(i=1;i<=12;i++)|
... etc.
```

The probability that in this approach no errors occur, the memory is sufficient, and thus the script is running tends to zero. As *JavaScript* only runs error-free or not at all and there are no error messages, a more promising method is chosen. For this, the entire *JavaScript* code of this clock is simply copied to the library. One possible approach:

First copy the clock routines from the source code of the HTML page beginning at line *function Zifferblatt()* to line *window.setInterval ("watch ()", 1000)* to the clipboard, then in *ESPBasic* load using the *FILE MANAGER* the formerly uploaded file from */uploads/bt93.js* and press *EDIT* to get to the editor. Insert the copied routine from the clipboard at the beginning or at the end. *SAVE* the modified file as new file: */uploads/bt93uhr.js*.

The BASIC listing now is short and small and there are no typos and/or memory problems. In the editor the listing is cleared by selecting and cutting all using the right mouse button. The displayed file name is to be deleted too, leaving ESPBasic work with the file */default.bas*. Its content is now the BASIC variant of the rest of the HTML page using a modified *JavaScript* file:

```
html |<html><body bgcolor="black"><br>|
html |<title>MSR mit Smartphone und Tablet</title>|
html |<canvas id="myCanvas" width=1280 height=720|
html |style="border:0px solid #c3c3c3"></canvas>|
html |<script src="/file?file=bt93uhr.js"
type="text/javascript"></script>|
</body></html>|
wait
```

A *SAVE* and *RUN* should display the clock as on the weburl
http://www.hjberndt.de/soft/BTUHR.html
in the browser, however, now locally and loaded from an *ESP8266* using *ESPBasic*!

Figure 137:
JavaScript-Basic-Clock on an
iPod-Touch2g using the local
network of the ESP

This script works fine for an elderly Safari browser using an iPod Touch

2nd generation.

Measuring values of passing time is a great thing, but the origin of this time data is NOT the ESP, but it is the time data of the caller, so the smartphone or tablet. Thus, after downloading the file using the script and the structure of the page, the ESP is obsolete, no longer needed. The clock continues to run in the browser, even if the ESP is turned off. *JavaScript* is running on a client.

3.2.7 ESPBASIC: BASIC AND JAVASCRIPT

In order to obtain measured data from the ESP board using *ESP8266Basic*, this language provides e.g. io routines, to measure and display values in the form of meters and textboxes. If the data is to represented by JavaScript, an interface between the two languages is necessary to access these values, even after the HTML page is set up complete. Using the two *ESPBasic* functions *HTMLID* and *HTMLVAR* there is a tunnel between the two languages. To illustrate, the value of the *ESPBasic* variable *x* is to be displayed by *JavaScript*. To get this going *x* is bound to an HTML element of *ESPBasic*, a *meter*. Immediately after the display of the *meter* the ID of the HTLM element is obtained calling the function *HTMLid()*, to access the meter element through its unique identifier using *JavaScript* calling there the function *document.getElementByld()*.

```
x = 99
meter x,0,100
id = htmlid()
html |<script>document.write(Date());|
html |var name_element = document.getElementById("| &
htmlvar(id) & |");|
html |var name =
name_element.value;document.write(name);|
html |</script>|
print id
end
```

The above example prints next to the *meter* element the current date in *JavaScript* and then, without any space, the set value of *x*, 99. The *JavaScript* variable *name_element* is assigned corresponding to the ID in BASIC and was requested by *HTMLid()*. This HTML element has a value that is obtained using *name_element.value()*. Thus, the *JavaScript* variable *name* has the value 99, which then appears calling *document.write()* next to the date. At the end of the print statement the ID is output, to see it change by every run. In this short example the page is set up only once, so the call to *HTMLvar()* could be omitted because BASIC writes the ID only once. Usually source of the generated HTML page is helpful to view and debug if getting issues.

3.2.8 ESPBASIC: OSCILLOSCOPE

Using the above technique the exchange of values between BASIC and *JavaScript* a graphical display of measured values by the use of existing libraries should be possible. For a test the library *bt93.js* is employed to do a presentation "NoTrigger" as on:

http://hjberndt.de/soft/BTLAUF.html

Using the HTML-source of that page, another colour design, but mostly same source code, this script creates a fast data representation of the analogue input from ESP8266F with its attached LDR. The 100 ms reflection of the oscillating red LED in the measurement routine is clearly visible. After a short while the daylight is registered after rotating the assembly.

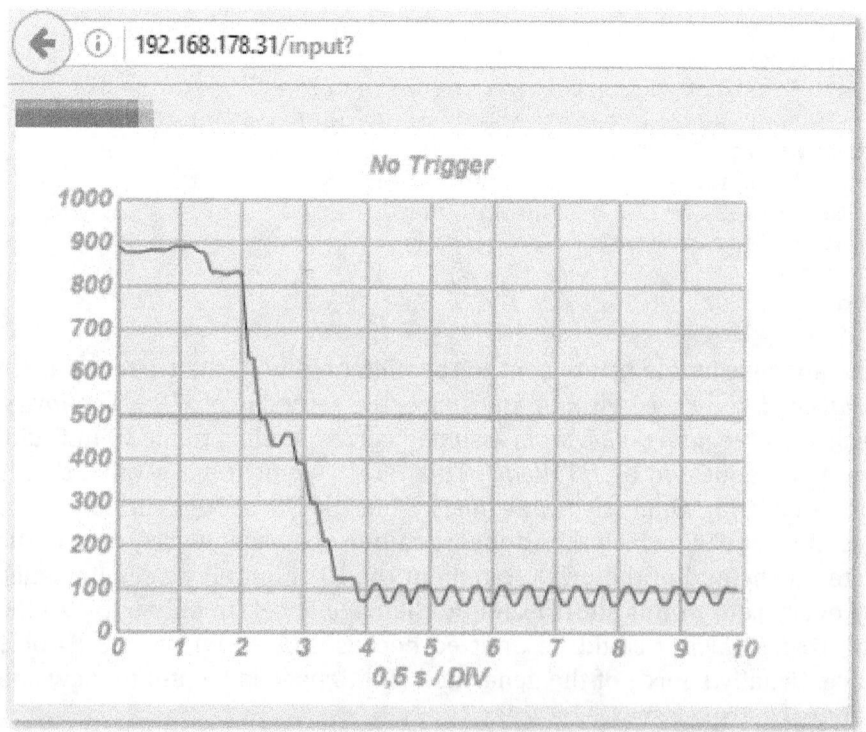

Figure 138:
Fast measurement data in a browser using ESPBasic and JavaScript

This adaptation embeds HTML/*JavaScript* source code in BASIC completely and technically this just fits in memory doing like that. For memory reasons spaces are omitted resulting in a badly readable listing.

```
x = 0
y = 88'NoTrigger
meter y,0,1000
idy = htmlid()
timer 100,[messen]
html |<html><body bgcolor="whitesmoke"><br>|
html |<canvas id="myCanvas" width=480 height=320
style="border:0px solid #c3c3c3"></canvas>|
html |<script src="/file?file=bt93.js"
type="text/javascript"></script><script
type="text/javascript">|
html |var name_element = document.getElementById("| &
htmlvar(idy) & |");|
html |const NMAX=100;var filled=false;var tmax=10;|
html |var TabY = new Array(NMAX);var TabX = new Ar-
ray(NMAX);|
html |function Messen(){var i,x,y,t,now=new Date();|
html |if(filled) {y=TabY[NMAX-1];   for(i=NMAX-1;i>0;i--)
TabY[i]=TabY[i-1]; TabY[0]=name_element.value;}|
html |else {for(i=0;i<NMAX;i++)  {x=i/NMAX*tmax;
=name_element.value;TabX[i]=x;|
html |TabY[i]=y} filled=true;}}|
html |function Anzeigen(){var i; Grafik(AN); Messen();
yachse="";xachse="0,5 s / DIV";titel="No Trigger";|
html |farbe ="White" ;cls(); farbe="whitesmoke"; hinter-
grund(); farbe="DarkGrey" ; Diagramm(0,tmax,0,0,1000,0);
farbe="gray"; |
html |DiaLinie(0,500,tmax,500); farbe="black "; |
html |for(i=0;i<NMAX-1;i++) DiaL-
inie(TabX[i],TabY[i],TabX[i+1],TabY[i+1]);}|
html |window.setInterval("Anzeigen()",
50);</script><br></body></html>|
wait

[messen]
x =not x
io(po,15,x)
y = io(ai)
wait
```

The original source calculates a theoretically sine and fills the array. In the above example data acquisition is done by calling ESPBasics *io(ai)* in the measurement routine and assigning the value to the *meter* element, the handover of the id of this HTML element to *JavaScript*, which retrieves the Id by this value and finally using *name_element.value* to enter the data to *TabY [0]*, the front of the measuring array, which is drawn by the display routine.

The temporal representation of the curve is affected at three places in the listing. The routine *messen* is called every 100 ms by a timer, the chart using 10 divisions and *JavaScript* records and updates - or at least tries - the entire graph every 50 ms.

3.2.9 ESPBASIC: TY-PLOTTER

A TY-Plotter can be modelled for slower operations like this:

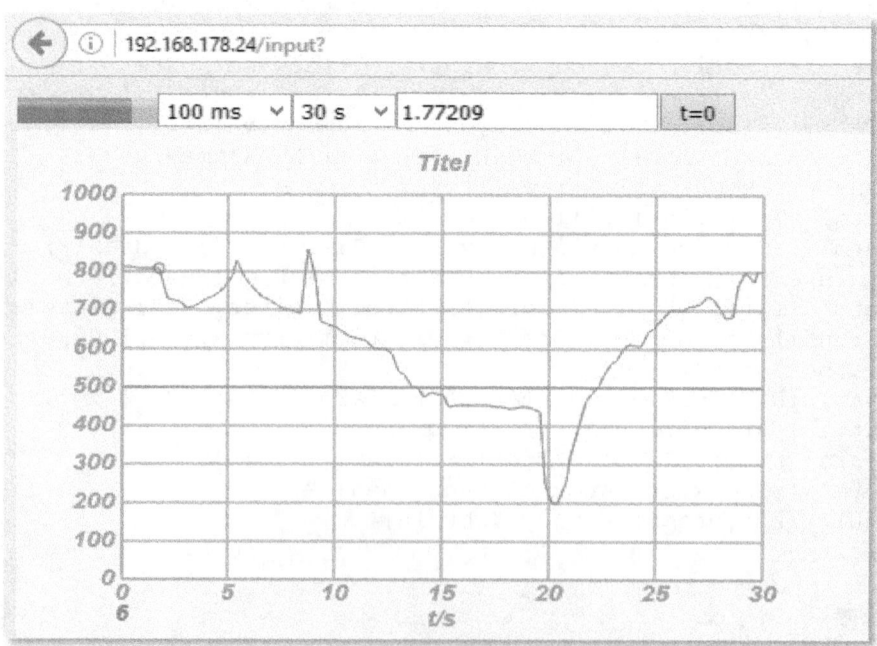

Figure 139: TY-Plotter for slow measurements

The example gets a measurement value every 0.1 second for 30 seconds and enters this data into the array capable of 100 values. The array index

- the 6 shown in blue - is calculated from duration and interval. The elapsed time is displayed in a *textbox*. For 100 measurement values and some user interface, this version quickly reaches memory limits.

Interval and measuring time can be freely set to apply different hardware and graphics capabilities of the client; it's an experimental playground only. To clear the chart on every run, the constant *NMAX* should be replaced by the variable *ix* as end value of the for-loop.

```
ti = "30 s"
td = "100 ms"
meter y,0,1000
idy = htmlid()
dropdown td, "dt, 10 ms, 20 ms, 30 ms, 50 ms, 100 ms, 200
ms, 500 ms, 1000 ms"
dropdown ti, "MBE, 10 s, 20 s, 30 s, 60 s, 120 s, 180 s,
600 s"
idti = htmlid()
t0 = millis()
textbox text
text = 0
idt = htmlid()
button "t=0", [reset]
timer 500 , [messen]
html |<html><body bgcolor="whitesmoke"><br>|
html |<canvas id="myCanvas" width=480 height=320
style="border:0px solid #c3c3c3"></canvas>|
html |<script src="/file?file=bt93.js"
type="text/javascript"></script><script
type="text/javascript">|
html |var meter = document.getElementById("| & html-
var(idy) & |");|
html |var tmaxs = document.getElementById("| & html-
var(idti) & |");|
html |var idt = document.getElementById("| & htmlvar(idt)
& |");|
html |const NMAX=100;var filled=false;var ix=0; var
tmax=100;var t,y;|
html |var TabY = new Array(NMAX+1);var TabX = new Ar-
ray(NMAX+1);|
html |function Messen(){|
html |t=idt.value; y=meter.value;ix =
(Math.round(NMAX*t/tmax));|
html |TabY[ix]=y;TabX[ix]=t;}|
```

```
html |function Anzeigen(){var i; Graf-
ik(AN);Messen();yachse="";xachse="t/s"; |
html |farbe ="GhostWhite " ;cls();farbe="white
";hintergrund(); farbe="DarkGrey" ; |
html |var res = tmaxs.value.substring(0,
3);tmax=parseInt(res); |
html |var step = 10;if(tmax<100)step=5;
       if(tmax<30)step=2;if(tmax<20)step=1;if(tmax>100)ste
p=0; |
html |Diagramm(0,tmax,step,0,1000,0); farbe="gray"; DiaL-
inie(0,500,tmax,500); |
html |farbe="blue";DiaText(0,-
120,ix);DiaPunkt(t,meter.value);farbe=ROT; |
html
|for(i=0;i<NMAX;i++)DiaLinie(TabX[i],TabY[i],TabX[i+1],Ta
bY[i+1]);} |
html |window.setInterval("Anzeigen()",
50);</script><br></body></html> |

[reset]
t0 = (millis()/1000)
wait

[messen]
y = io(ai)
if (text)>val(ti) then t0 = (millis()/1000)
text = (millis()/1000)-t0
timer val(td),[messen]
wait
```

Now the source code of that old TPU, which was initially rudimentary ported to the HTML5 canvas and *JavaScript* and there recycled for the first time, now can get an "H"-sign - as is common for German old-timer cars.

But now more recent sources are being used!

3.2.10 ESPBASIC: GAUGES AND JAVASCRIPT - SLOW

In an article found at:

http://www.esp8266.com/viewtopic.php?f=40&t=7056

Gauges and examples for the version 2 of ESPBasic are shared. The mechanism relies on the function ONLOAD. The advantage is that code will even run on oldest browsers, but - depending on the browser and graphics capability - each reload builds up the whole page completely and this may be an unattractive effect in faster measurements. Yet here the code for ESPBasic 2, in which the syntax sometimes deviates! Prerequisite is the UPLOAD of the *gauge.min.js* file to the ESPBasic file system.

```
bla = 99
ONLOAD [get.AI]
goto [show.page]

[get.AI]
ai bla
wait

[show.page]
wprint |<meta http-equiv="refresh" con-
tent="30;URL=/input?">|
wprint |<!doctype html>|
wprint |<html>|
wprint |<head>|
wprint |<title>Gtest</title>|
wprint |<script src=/file?file=gauge.min.js></script>|
wprint |</head>|
wprint |<body>|
wprint |<canvas id="gauge1"data-value=|
wprint htmlvar(bla)    ' 1. gauge paste here code
wprint | width="400" height="400"|
wprint |data-type="canv-gauge"|
wprint |data-title="Speed"|
wprint |data-min-value="0"|
wprint |data-max-value="1025"|
wprint |data-major-ticks="0 100 200 300 400 500 600 700
800 900 1000"|
wprint |data-minor-ticks="2"|
wprint |data-stroke-ticks="true"|
wprint |data-units="km/h"|
```

```
wprint |data-value-format="3.2"|
wprint |data-glow="true"|
wprint |data-animation-delay="10"|
wprint |data-animation-duration="200"|
wprint |data-animation-fn="bounce"|
wprint |data-colors-needle="#f00 #00f"|
wprint |data-highlights="0 30 #eee, 30 60 #ccc, 60 90
#aaa, 90 220 #eaa"|
wprint |></canvas>|
wprint |</body>|
wprint |</html>|
wait
```

gauge.min.js

The reference to these elements was found at:

https://github.com/Mikhus/canvas-gauges/wiki/Gauge-HTML-API

The image showing the three gauges humidity, temperature and bright-ness is another adaptation to the *ESPBasic*, but now in version 3.55. Be-cause of the three elements of the source code, which usually consists of HTML, is quite long, but works on the iPod Touch 2G. The readings are taken from a DHT11-Sensor as it has been used in the section on ESP8266Basic. This sensor only supplies integer values, but the meas-urement routine builds a moving average over 10 values from approxi-mately every second measures, achieving decimal places in the tempera-ture display. Random values can serve the experiment if there is no sen-sor available.

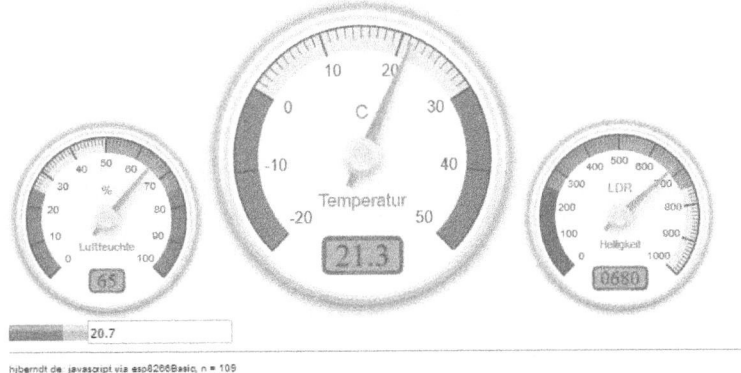

Figure 140: JavaSrcipt-Gauges using ONLOAD

```
dht.setup(11, 2)
dim y(10)
ix=1
n = 1
a = -1
timer 1000,[messen]
wprint |<!doctype html>|
wprint |<meta http-equiv="refresh" content="10;">|
wprint |<html>|
wprint |<head>|
wprint |<title>B3 Gauges</title>|
wprint |<script src=/file?file=gauge.min.js></script>|
wprint |</head>|
wprint |<body background="/file?file=b.png">|

wprint |<canvas id="gauge1" width="200" height="200"|
wprint |data-type="canv-gauge"|
wprint |data-title="%"|
wprint |data-min-value="0"|
wprint |data-max-value="100"|
wprint |data-major-ticks="0 10 20 30 40 50 60 70 80 90
100"|
wprint |data-minor-ticks="10"|
wprint |data-stroke-ticks="true"|
wprint |data-units="Luftfeuchte"|
wprint |data-value-format="2.0"|
wprint |data-glow="true"|
wprint |data-animation-delay="10"|
wprint |data-animation-duration="500"|
wprint |data-animation-fn="elastic"|
```

```
wprint |data-colors-needle="#e76 #f77"|
wprint |data-highlights="0 25 #fa0, 25 50 #eee, 50 75
#aaa, 75 100 #d31"|
wprint |data-onready="setInterval( function() {
Gauge.Collection.get('gauge1').setValue(| & htmlvar(h) &
|);}, 100);"|
wprint |></canvas>|
wprint |<script type="text/javascript">|
wprint |//Gauge.Collection.get('gauge1').setValue(88);|
wprint |</script>|
wprint |<canvas id="gauge2" width="320" height="320"|
wprint |data-type="canv-gauge"|
wprint |data-value-format="2.1"|
wprint |data-type="canv-gauge"|
wprint |data-colors-needle="#e76 #f77"|
wprint |data-highlights="-20 0 #fa0, 0 30 #eee, 30 50
#d31"|
wprint |data-title="C"|
wprint |data-units="Temperatur"|
wprint |data-min-value="-20"|
wprint |data-max-value="50"|
wprint |data-major-ticks="-20 -10 0 10 20 30 40 50"|
wprint |data-minor-ticks="10"|
wprint |data-stroke-ticks="true"|
wprint |data-animation-fn="elastic"|
wprint |data-onready="setInterval( function() {
Gauge.Collection.get('gauge2').setValue(| & htmlvar(t) &
|);}, 100);"|
wprint |></canvas>|
wprint |<canvas id="gauge3" width="200" height="200"|
wprint |data-type="canv-gauge"|
wprint |data-title="LDR"|
wprint |data-min-value="0"|
wprint |data-max-value="1000"|
wprint |data-major-ticks="0 100 200 300 400 500 600 700
800 900 1000"|
wprint |data-minor-ticks="10"|
wprint |data-stroke-ticks="true"|
wprint |data-units="Helligkeit"|
wprint |data-value-format="4.0"|
wprint |data-glow="true"|
wprint |data-animation-delay="10"|
wprint |data-animation-duration="500"|
wprint |data-animation-fn="elastic"|
wprint |data-colors-needle="#e76 #f77"|
```

```
wprint |data-highlights="0 250 #555, 250 500 #999, 500
750 #aaa, 750 1000 #eee"|
wprint |data-onready="setInterval( function() {
Gauge.Collection.get('gauge3').setValue(| & htmlvar(a) &
|);}, 100);"|
wprint |></canvas><br>|
meter h, 0,100
textbox h
meter t,-20,50
textbox t
meter a, 0,1000
textbox a
wprint |<hr noshade size=1><font size=1
face=Arial,Helvetica>hjberndt.de: javascript via
esp8266Basic, n = |& htmlvar(n) &|</font>|
wprint |</body>|
wprint |</html><br>|
mem = ramfree()
wprint htmlvar(mem) & " " &htmlvar(n)
wait
[messen]
t = 22 + rnd(10) - 5
h = 50 + rnd(50) -25
t = DHT.TEMP()
h = DHT.HUM()
i = DHT.HEATINDEX()
if t = 0 then t = 22 + rnd(10) - 5
if h = 0 then h = 50 + rnd(50) -25
a = io(ai)
y(ix) = t
ix = ix + 1
if ix>10 then ix = 1
mw = 0
for i = 1 to 10
 mw = mw + y(i)
next i
mw = mw/10
n = n + 1
if n>10 then t = mw
serialprintln h
serialprintln t
wait
```

3.2.11 ESPBASIC: *Gauges and Javascript - Fast*

A full speed display using these *JavaScript* elements only will be achieved without a new screen rendering on every new measurement. This method has been used above showing the charts using *bt93.js*. A single gauge element for a rapidly changing brightness looks like this, altering the original code as little as possible:

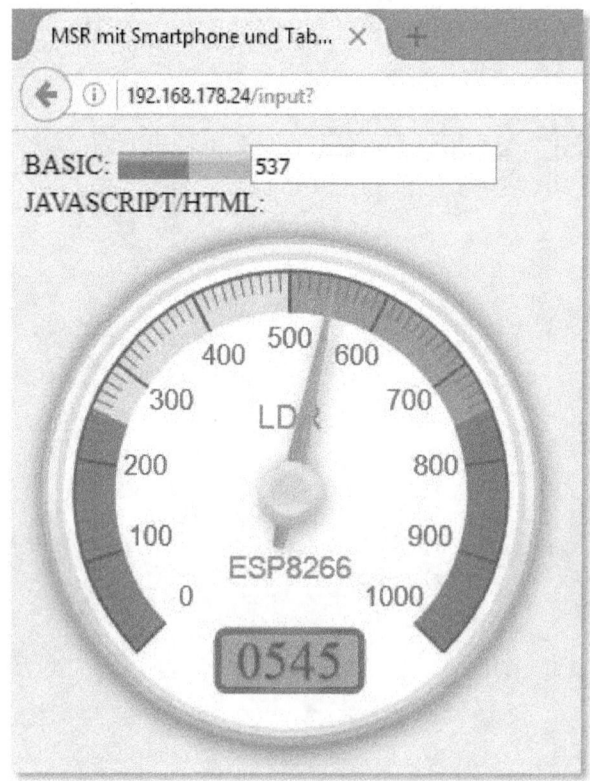

Figure 141:
Fast 5 mps realtime measurements in a browser using JavaScript

```
html "BASIC: "
x = 0
meter h,0,1000
idh = htmlid()
Textbox h
html "<br>JAVASCRIPT/HTML:<br>"
timer 100, [messen]
wprint |<!doctype html>|
wprint |<html>|
```

```
wprint |<head>|
wprint |<title>MSR mit Smartphone und Tablet</title>|
wprint |<script src=/file?file=gauge.min.js></script>|
wprint |</head>|
wprint |<body bgcolor="whitesmoke">|
wprint |<canvas id="gauge1" width="320" height="320"|
wprint |data-type="canv-gauge"|
wprint |data-title="LDR"|
wprint |data-min-value="0"|
wprint |data-max-value="1000"|
wprint |data-major-ticks="0 100 200 300 400 500 600 700
800 900 1000"|
wprint |data-minor-ticks="10"|
wprint |data-stroke-ticks="true"|
wprint |data-units="ESP8266"|
wprint |data-value-format="4.0"|
wprint |data-glow="true"|
wprint |data-animation-delay="0"|
wprint |data-animation-duration="0"|
wprint |data-animation-fn="linear"|
wprint |data-colors-needle="#e76 #f77"|
wprint |data-highlights="0 250 #fa0, 250 500 #eee, 500
750 #aaa, 750 1000 #d31"|
wprint |data-onready="setInterval( function()
{Gauge.Collection.get('gauge1').setValue( show());},
100);"|
wprint |></canvas>|
html |<script>function show(){var meter= docu-
ment.getElementById("| & idh & |"); return (me-
ter.value);}</script>|
wprint |</body>|
wprint |</html>|
wait

[messen]
h = io(ai)
x = not x
io(po,15,x)
wait
```

The inner function *show*() returns the brightness value to the gauge. The timer invokes every 0.1 s the measurement routine, whereby the variable *h* changes and thus the value of the *meter* element. It can be seen that BASIC is a little straighter using its simple *meter* element. The gauge has a number of settings for animation and these might slow down rapid display changes. Presumably, such a display for such fast processes is not appropriate at all. The flashing red LED at a frequency of 5 Hz and the resulting brightness can be followed by the instrument well. LDR and ADC need some time either.

3.2.12 ESPBASIC: ON A FOREIGN PC

The *ESP8266* allows very simple mobile measurements using the smartphone or tablet are easy to perform. But fast and big foreign desktop computers can be connected to the small wireless measurement and control system, since nothing needs to be installed or copied. If the foreign system even allows connections via Wi-Fi, all needed is a running browser on the PC. If the remote system only has network access via LAN, a smartphone can be used as an intermediary.

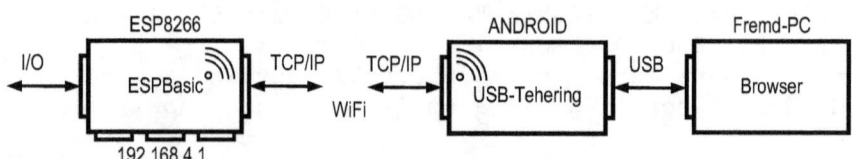

Figure 142: ESP8266 on foreign computer via USB tethering

In addition to a Wi-Fi hotspot, a smartphone can perform its tethering as well as by Bluetooth as by USB. In most cases, desktop PCs come with a free USB port for using memory sticks.

Once the USB connection is made, Windows typically searches for device drivers, depending on what device the smartphone can be recognized as via USB. With USB tethering this may take a while if not even a little longer. After that while another network connection should be available, possibly without Internet access, if the mobile data on the phone is off and only Wi-Fi is turned on. In order to implement measurements in the browser of the third-party computer via *ESP8266*, the following steps can lead to success:

- Connect ESP8266 in ESPBasic-mode to a power source (USB)
- Connect smartphone via WLAN to the ESP hotspot (AP)
- Connect smartphone via USB to the external computer
- Smartphone: Activate USB-Tethering
- Wait for another network connection on the PC (Taskbar)
- Start the browser and call 192.168.4.1

Usual administrative configurations then allow the mobile ESPBasic measuring system to be used on a remote computer; only the screen size may need to be adjusted using canvas output. Here is an example on a well-protected office machine:

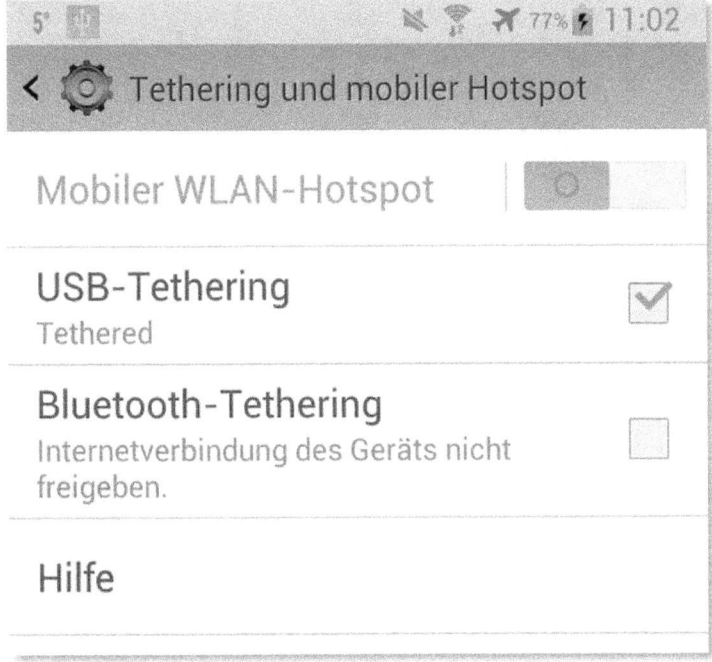

Figure 143: Via USB to the browser on a remote system

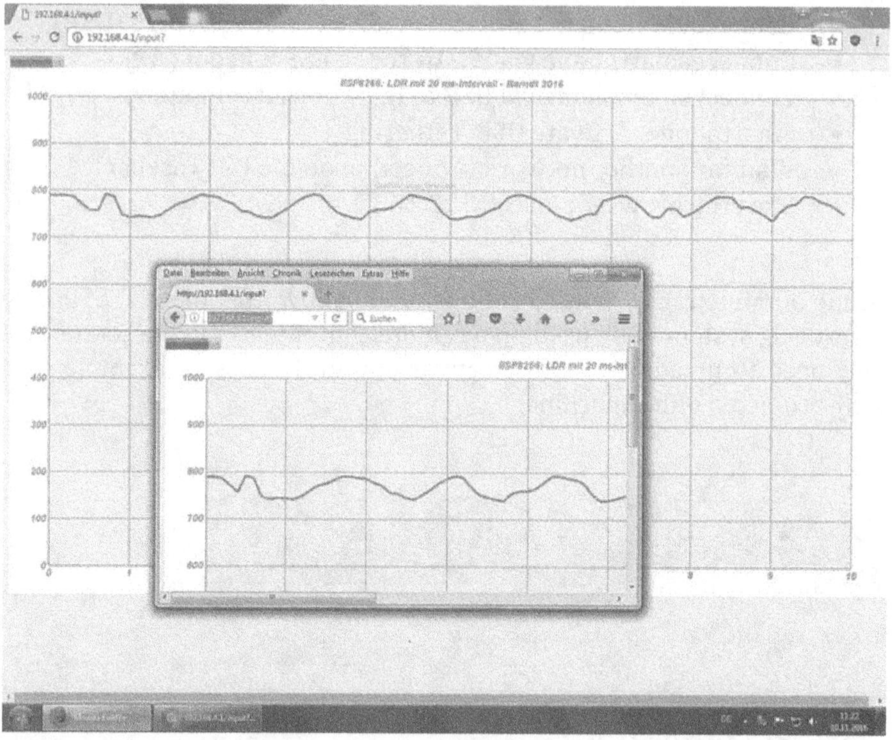

Figure 144: Firefox and Chrome in parallel on an office PC as an ESP8266 measuring platform

3.2.13 ESPBASIC: ADS1115 4-CHANNEL-ANALOGUE-INPUT 16 BIT

On the ESP8266 there is one 10-bit analogue converter, an ADC. For more accurate measurements or just more analogue input channels, an extension is required. A widespread and cheap chip is the ADS1115, providing 4 analogue input channels in a 16-bit resolution. In addition there is a PGA, a programmable amplifier, allowing even measurements in the microvolt range. Another highlight is the ability to program the inputs as a differential amplifier. The result is a universal mobile programmable wireless measurement system accessable by a browser.

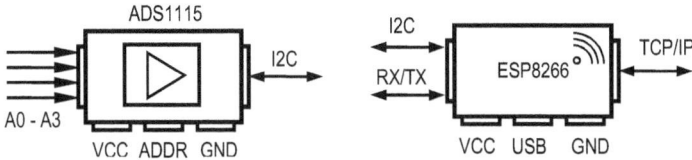

Figure 145: ESP and 4 analogue input channels using ADS1115

Figure 146:
ADS1115 as
breakout

The connection of the two components is done by the two I²C lines SCL/SDA. On the ESP by default the connections are GP02 (SCL) and GP04 (SDA). ADDR on the ADS1115 board is connected to ground (GND). Using an ESP *Witty Cloud* and USB powering the supply voltage can be taken from there.

ADS1115 using I2C and ESPBASIC

Getting started with the ESP and the ADS is done best first in the Arduino IDE and the corresponding library of Adafruit. There the different amplifier modes of the breakout are illustrated very clearly:

https://github.com/adafruit/Adafruit_ADS1X15

```
// The ADC input range (or gain) can be changed via the following
// functions, but be careful never to exceed VDD +0.3V max, or to
// exceed the upper and lower limits if you adjust the input range!
// Setting these values incorrectly may destroy your ADC!
//
ADS1115
//        ------
// ads.setGain(GAIN_TWOTHIRDS);
// 2/3x gain +/- 6.144V  1 bit = 0.1875mV (default)
// ads.setGain(GAIN_ONE);
// 1x gain   +/- 4.096V  1 bit = 0.125mV
// ads.setGain(GAIN_TWO);
// 2x gain   +/- 2.048V  1 bit = 0.0625mV
// ads.setGain(GAIN_FOUR);
// 4x gain   +/- 1.024V  1 bit = 0.03125mV
// ads.setGain(GAIN_EIGHT);
// 8x gain   +/- 0.512V  1 bit = 0.015625mV
// ads.setGain(GAIN_SIXTEEN);
// 16x gain  +/- 0.256V  1 bit = 0.0078125mV
```

After successful testing using this library and its functions and analysing the data sheet of the ADS1115 the BASIC programming can be implemented and continue without the Adruino IDE. Since *ESPBASIC8266* has a set of I²C commands, the port to this BASIC is not that difficult. The configuration of the ADC using the I²C address with the decimal value of 72 (0x48) in a subroutine [*conf*] and the variable *config* holding the correct value, is done like this:

```
[conf]
adr = 72
h = config/256
l = config and 255
i2c.begin(adr)
i2c.write(1)
i2c.write(h)
i2c.write(l)
i2c.end()
'delay 8
return
```

The voltage in mV in the variable *u* at the input specified by *config* is obtained by calling this subroutine:

```
[read]
gosub [config]
i2c.begin(adr)
i2c.write(0)
i2c.end()
i2c.requestfrom(adr,2)
h = i2c.read()
l = i2c.read()
hl = h * 256 + l
u = hl * 0.1875
return
```

To measure the voltages at the inputs A0 and A1, the configuration must be altered respectively. The following *ESPBASIC* section performs a measurement of the two input lines A0 and A1 on the ADS and supplies the voltage in mV in the variables *u1* and *u2:*

```
[messen]
conf = 49539
gosub [read]
u1 = u
config = 53635
gosub [read]
u2 = u
wait
```

The values for the configuration are obtained by analysing the ADS data sheet and the Adafruit library.

3.2.14 ESPBASIC: 2-CHANNEL-MEASUREMENT

An application of the *ESP8266* and the ADS1115 using *ESPBASIC*, HTML and *JavaScript* is the loading and unloading curve of a resistor-capacitor combination. It is aimed to be representative of other TY-measurements with several channels, which can be done by use of a browser and the intention to produce corresponding diagrams or charts directly.

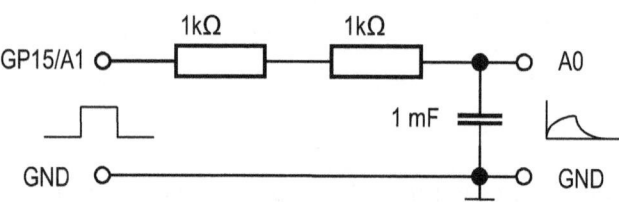

Figure 147: Measuring circuit for 2-channel measurement at the A0/A1

The digital output GP15 with the red LED of the *Witty Cloud* ESP switches the board voltage of 3.3 Volts, whereby the capacitor can recharge. Accordingly, the capacitor discharges through the resistors when GP15 outputs 0 Volts. In combination of the *ESPBASIC* techniques, as explained in section 2 of this book, controlling the LED/GP15 is done by the user by a button in the browser window.

The overall listing again is a "MashUp" of *ESPBASIC*, HTML and *JavaScript*. The library *bt93.js* is in the storage area of the *ESPBasic* and provides the graphing of the chart. BASIC handles the user interface and I²C control. The timer-controlled measurement runs in BASIC, while the graph is updated by *JavaScript*, as seen elsewhere in this section. The following lines represent the overall program listing with the fictive name *adsRCbutton.bas* which can be typed as text in the BASIC editor in this order, without the non-Basic text passages - of course.

First the screen is drawn displaying the button and displays for measured values.

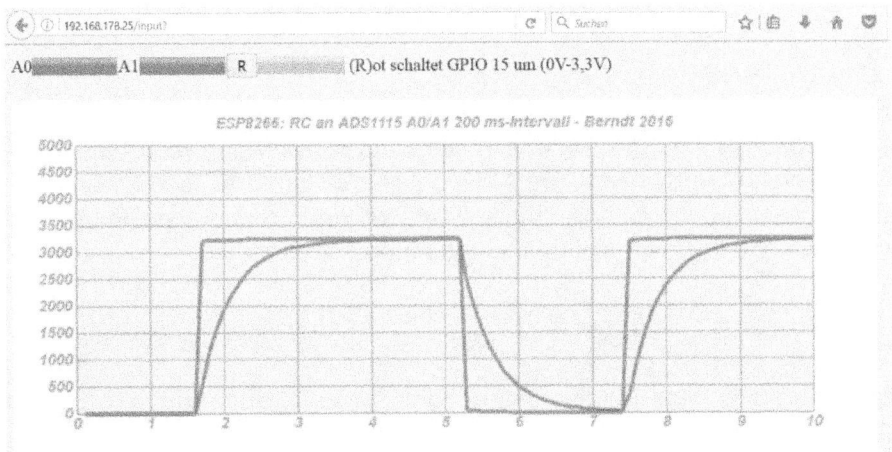

Figure 148:
2-channel measurement in Firefox in real time; BASIC, HTML, JavaScript

```
'rc entladekurve 2 kanal mit button gpio15 an ads1115
x = 0
y = 88 'btB3NoTrigger<<
html |A0|
meter y,0,3300
idy = htmlid()
html |A1|
meter y1,0,3300
idy1 = htmlid()
button "R",[rot]
meter r, -1,0
html | (R)ot schaltet GPIO 15 um (0V-3,3V)<br>|
rot=-1
```

Next the HTML5-canvas is created:

```
html |<html><body bgcolor="whitesmoke"><br>|
html |<canvas id="myCanvas" width=800
height=320</canvas>|
```

Here is the place to alter width and height of the canvas - the graphic area. The background colour in whitesmoke is not recognized by every browser, but is merely a gimmick. The following lines include the *JavaScript* library *bt93.js* in the *ESPBASIC*-specific way. This file should have been previously uploaded into the BASIC memory area.

```
html |<script src="/file?file=bt93.js"
type="text/javascript"></script>|
```

Next *JavaScript* starts with the definition of the variables. Using *HTML-Var JavaScript* gets the IDs of the elements *meter* y and *meter* y1 to access its value. In this way, the ADS values are shared to the *JavaScript* in the browser of the calling Phones/Tablets. Then fixing a setting of 100 values in two arrays for both analogue inputs are applied.

```
html |<script type="text/javascript">|
html |var name_element = document.getElementById("| &
htmlvar(idy) & |");|
html |var name_element1 = document.getElementById("| &
htmlvar(idy1) & |");|
html |const NMAX=100;var filled=false;var tmax=10;|
html |var TabY = new Array(NMAX);var TabX = new Ar-
ray(NMAX);var TabY1 = new Array(NMAX);|
```

The routine *Messen()* fills the data arrays. Moving the individual values ensures an impression of moving the measurement curve from right to left through the graph. The current measurement is entered into the arrays last element by *name_element.value*.

```
html |function Messen(){var i,x,y;|
html |for(i=1;i<NMAX;i++){TabY[i-1]=TabY[i];   TabY1[i-
1]=TabY1[i];TabX[i]=i/NMAX*tmax;}|
html |TabX[NMAX-1]=tmax;TabY[NMAX-
1]=name_element.value;TabY1[NMAX-
1]=name_element1.value;}|
```

Both vertical lines "|" or characters of an *html* statement must be on the same line with no line break in real Basic-Code.

The routine *Anzeigen()* initializes the graphic using the *bt93.js* library and calls one measurement, prepares and draws two measurement curves using the specified colours, represented by a line width of 3, into the chart.

```
html |function Anzeigen(){var i;  Graf-
ik(AN);Messen();yachse="";xachse="";titel="ESP8266: RC an
ADS1115 A0/A1 200 ms-Intervall - Berndt 2016";|
```

```
html  |farbe ="White"
;cls();farbe="whitesmoke";hintergrund(); farbe="darkgrey"
; Diagramm(0,tmax,0,0,5000,0); farbe="gray";  |
html  |ctx.lineWidth=3;   |
html  |for(i=0;i<NMAX-
1;i++){farbe=BLAU;DiaLinie(TabX[i],TabY[i],TabX[i+1],TabY
[i+1]);|
html
|farbe=ROT;DiaLinie(TabX[i],TabY1[i],TabX[i+1],TabY1[i+1]
);}}|
```

At the end of the initialization of the Web page in the user's browser, *JavaScript* calls the routine *Anzeigen()* once every 500 ms. This ends the script, the body and the HTML.

ESPBASIC takes over using its own time at a 500 ms interval, to initiate every 0.5 seconds a measurement too.

```
html  |window.setInterval("Anzeigen()",
500);</script><br></body></html>|

timer 500,[messen]
wait
```

The following measurement routines - explained generally above - now are used again, except for some minor variable name changes.

[messen]
```
config = 49539
gosub [read]
y = u
config = 53635
gosub [read]
y1 = u
r = io(laststat,15)
wait
```

[conf]
```
adr = 72
h = config / 256
l = config and 255
i2c.begin(adr)
i2c.write(1)
```

```
i2c.write(h)
i2c.write(l)
i2c.end()
'delay 8
return
```

[read]
```
gosub [conf]
i2c.begin(adr)
i2c.write(0)
i2c.end()
i2c.requestfrom(adr,2)
h = i2c.read()
l = i2c.read()
hl = h * 256 + l
u = hl * 0.1875
return
```

[rot]
```
io(po,15,rot)
rot = not rot
wait
```

At the end of this listing the routine [*rot*] is associated to the red button for switching the LED and GP15 to charge and discharge the capacitor by the user showing the corresponding curve in real time from/on different phones/tablets simultaneously.

The ESP8266 system using *ESPBASIC* represents an absolutely independent and extremely inexpensive measuring and control system, which is programmable and can be used without additional software or hardware in any current browser. The hardware includes a Wi-Fi hotspot and if the on-board 10bit ADC is insufficient, a low-cost expansion - as shown above - can be used to perform almost professional and complex measurement.

3.3 CHINESE WHISPERS: LANGUAGE LOOP

Figure 149: Chinese whispers using Android and Windows

3.3.1 *RFO-TCP/IP-NETCAT-VBS*

In [1], the on-board sensors were selected by voice using a short *rfo-*Basic program and Google voice recognition on an Android Smartphone - no problem. In this little application a voice loop similar to the game 'Chinese whispers' is built.

Some words are spoken into the microphone of an Android smartphone, recognized text by Google is sent via TCP/IP to a Windows tablet by *NetCat*, there a *VBScript* using text to speech sends this text to the tablets speaker or any other audio output. The smartphone again recognizes this text as the first spoken phrase (or not) and the loop repeats. Depending on the microphone distance, the detection can change.

On the part of the Android device an *rfo*-Basic program logs on to the Windows-Tablet at port 34 via TCP/IP and does the Google Voice recognition. *NetCat* on Windows is waiting and executes *cmd.exe* at a connection and routes all data to the caller. *Rfo*-BASIC now can control Windows by command line prompt.

On the Windows machine, a script for voice output is invoked using the next lines, saved as *spk.vbs* on the desktop.

```
Set Sapi = Wscript.CreateObject("SAPI.SpVoice")
Set Args = WScript.Arguments
For i= 0 To args.Count-1
     a=args(i)
     a=replace (a,"_"," ")
     wscript.sleep 2000
     Sapi.speak a
next
```

This script is speaking text passed as an argument. Because arguments are separated by spaces on the command line, Android sends an underscore instead of a space. This is reversed in this script, not interpreting every word as a new argument and jolting the speech. The two second delay is to restart Google speech recognition on the Android. Using the example shown in the section on *NetCat* Android gets access to Windows. On the desktop, there is another file called *ncCmd.bat* with the following content:

```
C:\Temp\nc\nc -L -p 34 -e cmd.exe
```

The explanation for this line is found in chapter *Software Elements/ NetCat*. On the Android smartphone or tablet an *rfo*Basic program has to be run, looking like this:

```
SOCKET.CLIENT.CONNECT "192.168.1.102",34
SOCKET.CLIENT.STATUS r
IF r THEN
 PRINT "Connected to ";
 SOCKET.CLIENT.SERVER.IP a$
 PRINT a$+" Port 34"
ELSE
 END
ENDIF

FOR x =1 TO 5
 GOSUB speak
 a$="spk.vbs "+ theText$
 GOSUB key
```

```
NEXT x

SOCKET.CLIENT.CLOSE
END

key:
SOCKET.CLIENT.WRITE.LINE A$ +CHR$(10)+CHR$(13)
PAUSE 100
RETURN

REM Start of tts
speak:
STT.LISTEN
STT.RESULTS theList
s$=""
LIST.SIZE theList, theSize
FOR k= 1 TO theSize
 LIST.GET theList,k,theText$
 PRINT theText$
 rem tts.speak thetext$
 IF k=1 THEN first$=theText$
 s$=s$+thetext$
NEXT k
thetext$=first$
tts.speak thetext$
CLIPBOARD.PUT s$
RETURN
```

Rfo-BASIC was the main topic in eBook [1] and will only be applied, here a brief description:

The first BASIC section calls the Windows tablet using its local IP on port 34 via TCP/IP. There *NetCat* executes *cmd.exe*. Next, five times the speech input and recognition is called, spaces are replaced and using the subroutine *Key* the *VBscript* is called from the Windows command line.

To play the game, the steps look like this:

Windows: Run *ncCmd.bat*

Android: Run *tcp_ip_diktat StillePost.bas*
                  Speak at Google voice recognition prompt
                  some words

The distance microphone/speaker is best about 30 cm. After a while, the Windows Tablet "speaks" the recognized sentence or the Chinese Whispers first take (Speech language in Windows 8.1 is selected in speech recognition settings) and the delayed Google voice prompt should appear again for another record and recognition of the detected words.

3.4 ANDROID DICTED TO WORD (PAD)

3.4.1 RFO-TCP/IP-NETCAT-VBS

The "Chinese Whispers" above should be regarded as a pure gimmick for demonstrating the possibilities of interaction of the components. A little variation as dictation in WordPad might appear a little more useful. Based on a subjective feeling, the voice recognition works better on Android; the smartphone can act as a voice recorder for Windows subjectively more sophisticated word processing. Since not all PC got Microsoft Office installed, the sample is done using WordPad, available on any Win-system as a RTF-Editor.

Using *VBScript*, there is no direct access to TCP/IP and VBA in Office is not the subject of this book. Text entry here is done by the software method *SendKeys* as controlled keystrokes, as in the examples on *VBScript* in this book in the second section. On the Windows Desktop, there is the *ncCmd.bat* file from above and a script called *Key.vbs* looking like this:

```
Set Args = WScript.Arguments
set WshShell = WScript.CreateObject("WScript.Shell")
For i= 0 To args.Count-1
     WScript.Sleep 100
     a=args(i)
     a=replace (a,"_"," ")
     WshShell.SendKeys a
next
```

The strings sent by Android include replaced spaces as underscores which have to be undone in this script. On Android, a *rfo*-BASIC program is running and connects and starts *WordPad* by the command line. The script *Key.vbs* is called using *Hello World* as an argument by command line of Windows. Finally WordPad is stopped without storing these beautiful words and the connection is closed.

```
SOCKET.CLIENT.CONNECT "192.168.1.102",34
SOCKET.CLIENT.STATUS r
IF r THEN
 PRINT "Connected to ";
```

```
 SOCKET.CLIENT.SERVER.IP a$
 PRINT a$+" Port "+port$
ELSE
 END
ENDIF

REM Ab hier steht die Verbindung

a$="WordPad.exe"
GOSUB key

a$="key.vbs Hallo Welt"
a$=replace$ (a$," ","_")
GOSUB key
pause 2000

a$="taskkill /IM WordPad.exe"
GOSUB key
a$="key.vbs %n"
GOSUB key
PAUSE 1000

SOCKET.CLIENT.CLOSE
END

key:
SOCKET.CLIENT.WRITE.LINE A$ +CHR$(10)+CHR$(13)
PAUSE 100
RETURN
```

3.5   CIRCULAR TRAFFIC

3.5.1   NETCAT/WLAN/SERIALTRANSFER/HC06/FTDI/REALTERM

This composition is a connection loop using some of the components discussed in this book. In turn, it is intended to show possibilities for own problem solving, whereby only the actual connection and its function is applied and checked.

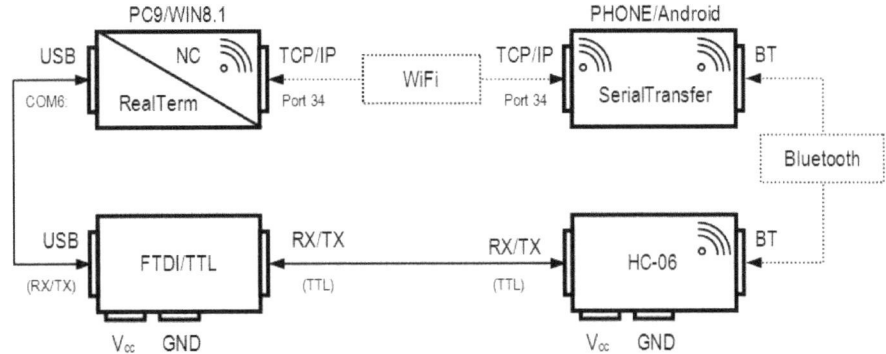

Figure 150: Looped data connection using software and hardware

The figure shows the loop starting at the upper left using a Windows tablet running *NetCat* to get access via TCP/IP or Wi-Fi. An Android smartphone using the PlayStore app *SerialTransfer* accepts the connection via TCP/IP and redirects data wireless using this software as a Bluetooth connection to the *HC06*-Bluetooth module, which is connected to a *FTDI-adapter* using its two wires RX/TX, passing data to another Windows tablet/PC by USB. The incoming information or data is picked up there by a COM interface using *RealTerm* or will be forwarded.

The result is a multiple control of Windows with all options of the command line - at the same time - from several different devices. A slightly more detailed explanation should allow the reproducibility and replica.

On a Windows Tablet there is *NetCat* using the execute parameters, as described elsewhere in this book. The *c:\temp\nc\nc -L -p 34 -e cmd.exe* line fulfils this purpose. The tablet can then be reached at port 34 and serves the caller by execution of the specified executable following the *-e* argument with its standard output. On the smartphone SerialTransfer provides using its TCP/IP branch to connect to that listening tablet using an assumed local IP of 192.168.1.100 on port 34.

The Android app (similar to a part of *RealTerm* on Windows) mirrors the data to a paired and connected Bluetooth device. Without additional software, or using hardware only, the data flows through the *HC06* and by wire to the *FTDI-Adapter* as an USB-connection seen by Windows as e.g. COM6. Any Windows software supporting RS232 devices can here pick up the communication - without the hard to reproduce Bluetooth issues of Windows. In this example the RS232-software *RealTerm* will take over. The loop gets started like this:

1. Run *NetCat* on Windows tablet
2. Connect *HC06* to *FTDI-adapter* until *HC06* flashes crosswise (RX/TX - TX/RX)
3. Run *SerialTransfer* app and connect to *HC06* (HC without flashing)
4. Connect *SerialTranfer* app to Windows tablet via TCP/IP
5. Connect *FTDI-adapter* to USB of the WinPC (i.e. COM6 comes up in device manager)
6. Run *RealTerm* and connect to port COM6
7. Disconnect and reconnect TCP/IP branch in *SerialTransfer* app

As a result, both in *RealTerm* at the end of the loop, as well as on Android in the middle, the Windows on the tablet is prompting to accept orders by the command line. As an example a directory tree can be output by typing *tree*.

3.6 OLD HARDWARE

3.6.1 RS232/FTDI/REALTERM/FTDI/BLUETOOTH

An old PC without network card and Bluetooth hardware serves an old RS232-interface or measuring device. For safety reasons this computer must do without appropriate hardware network capability, may be due to still using Windows XP. Nevertheless, there is demand for controlling this device by smartphone or tablet via Bluetooth and/or TCP/IP. This composition is built as a concrete solution without the need of any programming. The hardware is a PC running XP with its two USB ports. Alternatively, the RS223-device - here a *CompuLab* as used in [2] - can directly be connected to an available COM port, which then eliminates one USB connection. Many of the above explained components in this book are required to build this special solution.

The interface connects by its 9-pin socket directly or by an RS232-adapter to the old PC. On this computer *RealTerm* is installed and connects via, for example, COM5 with the external measuring system. Using this software it is possible to send and receive messages as ASCII or as hex commands. *RealTerm* mirrors the connection to an USB/TTLSerial-FTDI adapter to the other USB as e.g. COM10 logged to the old PC.

3.6.2 BLUETOOTH

An *HC06* module is connected using the RX/TX lines with TTL level of the *FTDI-adapter* in order to obtain Bluetooth connectivity. At this point, a wireless connection to the *CompuLab* is established and a smartphone or tablet using a Bluetooth terminal can access the device.

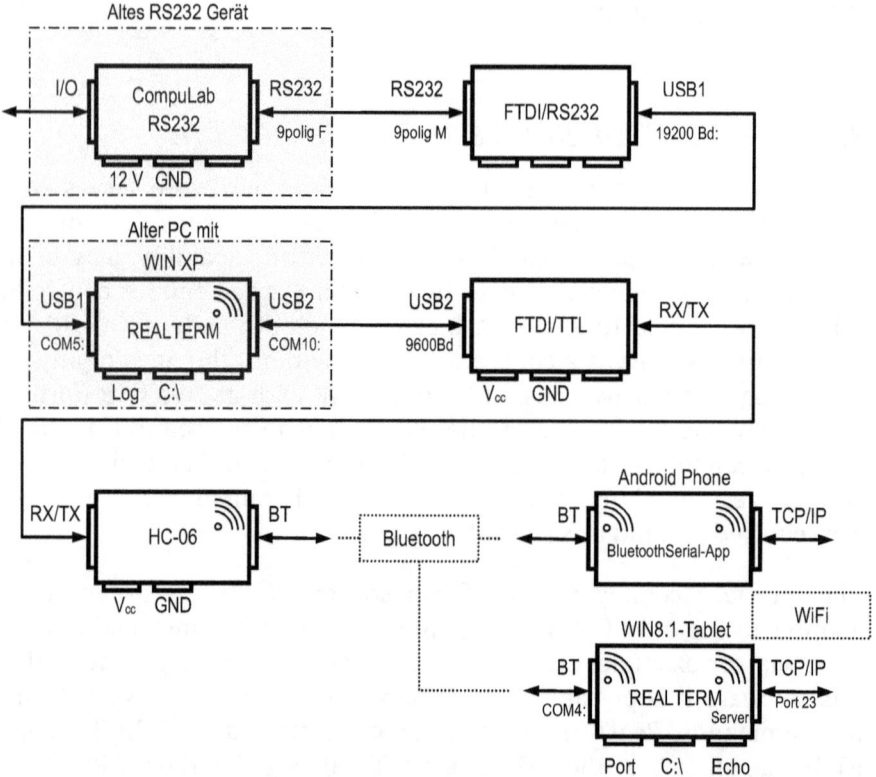

Figure 151: Old Windows PC and a measuring device is network-capable without hardware intervention

The Android app *SerialTransfer* of Next Prototypes accesses the Bluetooth connection and redirects to a network-capable smartphone or tablet as a TCP/IP connection. Thus, control of measurement hardware from any location is possible, if necessary.

CompuLab commands or queries can be found in the section *CompactDefinition* recognizes Digispark as Compulab. The decimal 211 (D4 Hex) supplies the state of the 8 digital inputs as a byte. 81 (51 Hex) to change the digital outputs to the next sent byte. The character string 5101 in hex mode turns on the least significant LED of the digital output. If all outputs are connected to the inputs, the query D3 sould return a value of 01.

This control may be issued at various points in this solution: in *RealTerm* running on the old PC as well as in the by Bluetooth associated *HC06* device. Because the original *CompuLab* is using 19200 baud, the incoming port of *RealTerm* is set to this transmission rate too. However, the echo port is operating at 9600 baud, the default speed of the *HC06* module.

3.6.3 *TCP/IP-WI-FI*

The Bluetooth connection on the Windows Tablet and an installed *RealTerm* is accessible via the corresponding COM5-interface assigned to the Bluetooth adapter as an incoming port with 9600 Baud of the *HC06* and mirrored to a telnet server on port 23 to serve client connections. The Android smartphone at the end of the line may connect to the tablet by its local IP, for example, PC9 to port 23. Now *CompuLab* or any other RS232 device on an old PC without Bluetooth and LAN/WLAN capabilities is accessible via TCP/IP. On the smartphone extensive measurements could be programmed using *rfo*-Basic, using the TCP/IP Commander simple controlling can be done without programming knowledge.

3.6.4 *TCP/IP-BLINK WITHOUT CODING*

The *Hello World* of light-emitting diodes is since Arduino the sketch Blink. With the presented solution a flashing LED can be realized using the smartphone and the *TCP Commander* of Next Prototypes (Android) repeated sending commands via TCP/IP to the Windows tablet where *RealTerm* via Bluetooth passes the command sequence to the *HC06* module, which serially speaks to the *FTDI-adapter* via USB and so to *RealTerm* on the old PC sending these bytes ultimately to the via USB/RS232-adapter connected *CompuLab* flashing a LED at one of its eight digital outputs.

3.7   JT65-PODCAST

3.7.1   VBS/WSJT (WINDOWS)

A live audio podcast reporting range and signal strength of weak and weakest HF signals of radio amateurs (HAM), generated automatically from the measurement data of the receiving device as well as from databases is the subject of the following composition. As a result there is information from and about the participants, as well as technical data by voice output.

A local shortwave receiver (or if there is no hardware: internet-based WebSDR) receives on the frequency 14076 kHz or 7076 kHz and demodulates the SSB signal into an audible tone sequence. The result sounds like a kind of melody with 65 or 9 notes, which is broadcast for about one minute from a long distance or a simple antenna at low power, according to the JT65 or JT9 method. The resulting audio is decoded by a free software WSJT-X on a Windows tablet, to get the call sign and other brief transmitted information. The software calculates a reception level based on the signal strength and displays it next to other information.

Although there is a free decoder for Android, free professional Windows software is used on the tablet in this composition runnung a lot better, than the more rudimentary Android copy, at least at the time this example was constructed.

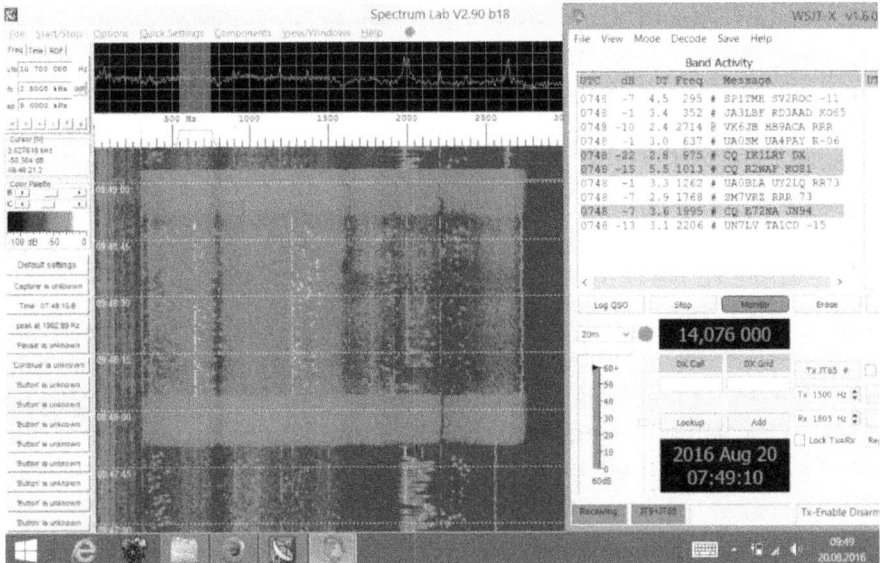

Figure 152: WSJT-X and Spectrum Lab with JT65/JT9 on the Windows Tablet doing 10 decodes a minute. Up to 16 calls were decoded per sequence

The software and the method come from K1JT - Joe Tayler (born 1941, Nobel Laureate Physics 1993 for astronomical studies). His method uses the properties of sound cards for decoding. Each transmission takes 46.8 seconds in which a maximum of 13 characters are transferred. The decoder requires exact PC time. As an example, the following data was displayed in the *Band Activity* window:

```
-21  0.7 1064 # CQ XT2AW IK92
```

The signal strength of this transmission corresponded to -21 dB, the PC clock was down by 0.7 seconds, the signal was in the audio spectrum at 1064 Hz, and the method was JT65 (#). A radio amateur with the call sign XT2AW from the locator IK92 called CQ.

Thus far, none of the treated components is needed or used.

Using a *VBScript* on this platform is to ensure that this additional information will be spoken. The decoder WSJT-X logs the data as text to the file *All.txt*, so the information can be processed later. In addition, name and country of origin should be announced acoustically as a kind of podcast. All this should, if possible, run without an Internet connection - outdoors, in the desert or in the mountains carrying radio and tablet - using a small local database that caches the call sign information at Internet availability in order to use the data offline. So to fill the empty local database with some data, an Internet connection is needed once.

3.7.2 CREATE DATABASE LOCALLY

Information about call signs from radio amateurs can be retrieved from the Internet. To avoid unnecessary burdens of these sources a local caching of the once found caller is done. Elsewhere in this book, the call sign query is explained using *VBS*cript. The result is a growing text file *JT65.TXT* in the *C:\Temp* folder and the approximately the following content:

...
CT1APP|Júlio Alexandre Martins da Silva|8005-265 FARO|Portugal
HB9CGH|MANFRED||Switzerland
9H1KR|Mario||Malta
EA3HKA|ELISEU|08551 TONA|Spain
...

First is the call sign followed by name, city and country, separated by a "|" character. This simple text file is used as a local database that is fed once online to work offline. The script first queries whether this file *C:\temp\jt65.txt* exists. If missing, it is created; else the script returns the number of items. Using the template *Verify that a file exists* the source of function *createJT65* is built like this:

```
' Verify that a File Exists
function createJT65
 Set objFSO = CreateObject("Scripting.FileSystemObject")
 If objFSO.FileExists(CALLFILE) Then
    Set objTextFile = objFSO.OpenTextFile(CALLFILE, 1)
    While Not objTextFile.AtEndOfStream
     strLine = objtextFile.ReadLine
     If inStr(strLine, "|") Then i=i+1
    wend
    WScript.Echo cstr(i)+" Einträge gefunden in
"+CALLFILE
 Else
```

```
    Wscript.Echo "Datenbank nicht gefunden. Erzeugt:
"&CALLFILE
    Set objFile = objFSO.CreateTextFile(CALLFILE)
 End If
End function
```

3.7.3 FILLING THE DATABASE

To add new call signs to the database CALLFILE - the global constant *JT65.TXT* - a corresponding subroutine is required. Using the example *Writing String Content to End of Existing File* the following code to *writeCallsign* has arisen:

```
' Writing String Content to End of Existing Text File
Sub writeCallsign(sign,name,city,country)
 Const FOR_APPENDING = 8
 strFileName = CALLFILE
 strContent  = sign+"|"+Name+"|"+City+"|"+Country+vbCrLf
 Set objFS = CreateObject("Scripting.FileSystemObject")
 Set objTS = objFS.OpenTextFile (strFile-
Name,FOR_APPENDING)
 objTS.Write strContent
End sub
```

3.7.4 DATABASE LOCAL QUERY

No database without a query. The by special characters separated entries of a call sign in the local database is provided by the function *findCallsign*, developed and based on the example *Read a Comma Separated Values Log*. The search starts at the beginning of the CALLFILE and ends if matching. The return value is the corresponding character string, or on failure, the unchanged call sign.

```
' Read a Comma Separated Values Log
Function findCallsign(sign)
 Const ForReading = 1
 findCallsign=sign
 Set objFSO = CreateObject("Scripting.FileSystemObject")
 Set objTextFile = objFSO.OpenTextFile(CALLFILE,
ForReading)
```

```
Do While objTextFile.AtEndOfStream <> True
  strLine = objtextFile.ReadLine
  If inStr(strLine, "|") Then
    sa = split(strLine, "|")
    If sa(0)=sign Then
     name=sa(1)
     City=sa(2)
     Country=sa(3)
     'findCallsign = name+" from "+city+", "+country
     If InStr(country,"Netherlands")>0 Then   country =
"The "+country
     findCallsign = name+" from "+""+country
     Exit function
    End if
  End If
 Loop
End function
```

3.7.5 SEARCH FOR CALLSIGN

The actual search for a received call sign performs a separate routine. If the call sign is found in the local database *GetCallSign* returns this information, if not found, an Internet search using *qrzcq.com* is done as described elsewhere in the section *VBScript*. If there is no Internet access, or if the search fails, *GetCallSign* delivers the transferred call sign without modification. At the beginning and at the end of the code there are some lines omitting unnecessary queries and manipulations to get an acceptable result for listening.

```
Function GetCallsign(sign)
 Const find="<p class=""haminfoaddress""><b style=""text-
shadow: 0px 1px 0px #f1f1f1, 0px 1px 3px #999; "">"
 GetCallsign=sign
 If sign ="DX"  Then Exit Function
 If sign ="RRR" Then Exit Function
 If sign ="GL"  Then Exit Function
 If sign ="73"  Then Exit Function
 If sign ="RST" Then Exit Function
 If sign ="PSE" Then Exit Function
 jt65 = findCallsign(sign)
 If jt65<>sign Then
  GetCallsign=jt65
  Exit Function
```

```
End If

strURL = "http://qrzcq.com/call/"+sign
Set objHTTP = CreateObject( "WinHttp.WinHttpRequest.5.1"
)
objHTTP.Open "GET", strURL:
On Error Resume next
objHTTP.Send
s= objHTTP.ResponseText:   ' WScript.Echo Len(s)
if objHTTP.Status = 200 Then
  ix=InStr(s,find)
  s=Mid(s,ix+Len(find),100)
  sa=Split(s,"<")
  If UBound(sa)>3 Then
    city=Mid(sa(3),6)
    country=Mid(sa(4),6)
    name=sa(0)
    writeCallsign sign,name,city,country
    WScript.Echo "New: ---------------------- " +sign+"
- "+name+" ----------------------"
    GetCallsign = name+" from "+""+country
  End if
End if
Set objHTTP = Nothing
End Function
```

3.7.6 USER DIRECTORIES

The simple local database-file *JT65.TXT* is hardcoded in the script as *C:\Temp\JT65.txt*. The log file is *All.txt*, without user intervention it can be found in a user directory such as:

```
C:\Users\...\AppData\Local\WSJT-X\ALL.TXT
```

These specific Windows directories are found by the sample script *ListItems in the Local Application Data folder*. With little adjustment the function *AllText* finds the log file in the original installation directory.

```
' List Items in the Local Application Data Folder
Function alltext
 Set objShell = CreateObject("Shell.Application")
 Set objFolder = objShell.Namespace(&H1c)
```

```
Set objFolderItem = objFolder.Self
alltext=objFolderItem.Path+"\WSJT-X\ALL.TXT"
End Function
```

3.7.7 MAIN CALL

The function or subroutine *callsign* is called once in the current minute at second 52, when the first present decoding is done. It will do the required job by reading the *All.txt* file and searching the current time (UTC). From there the routine tries to talk the data accordingly. The subroutine *timestamp* origins from the template *List the UTC time on a computer*.

One line of the log file will be examined and issued using the routines listed above to report the line acoustically accordingly. If the signal strength is less than -20 dB, then a customized output is created. If there is enough time left, the number of successful decodings in this minute is announced. If too many decodings exceeded the available time to report, a corresponding text can be heard.

The script is based on an English-language edition, but can work with appropriate text change using other languages.

```
1930
-10 Siegfried Reisch from Germany calls Gandler Rudolf from Austria with R-01
-12 Peter from Slovenia calls Jean-Pierre TACONNE from France with -13
-12 CQ from TOLY from European Russia
-13 Roger from England calls Neil Viskov from Estonia with I093
-10 Arif AKYOL from European Turkey calls Dumitru (Titi) Dobre from Romania with
 -09
-12 CQ from Alexander Volkov from European Russia
-11 CQ from MICHELE MANSI from Italy
 -7 André Dessibourg from Switzerland calls UB1ALQ with R-12
 -8 CQ from Nikolay Senkiv from Ukraine
-19 CQ from Wijnand Schwarte from The Netherlands, Overijssel
10 calls
```

Figure 153: JT65 as plaintext and sound by means of a script

The script *JT65 5g JT65TXT no city.vbs* as complete listing. Using *VbsEdit* this script can be compiled to an Windows-executable *JT65 5g JT65TXT no city.exe*.

```
'Const logfile="C:\Users\...\AppData\Local\WSJT-
X\ALL.TXT"
```

```
Const CALLFILE="C:\temp\JT65.txt"
WScript.Echo "This script produces speech. Use at your
own risk!"
WScript.Echo "JT-65 Decoder WSJT-X to speech callsign via
qrzcq.com."
WScript.Echo "Input log-file is: "& alltext
createJT65
WScript.Echo "Waiting for second 52 ..."


While True
 If Second(Time)>51 Then callsign
 WScript.Sleep 1000
Wend

' Read a Text File into an Array
'2158 -21   0.7 1064 # CQ XT2AW IK92
Sub callsign
 Set Sapi = Wscript.CreateObject("SAPI.SpVoice")
 Const fsys = "Scripting.FileSystemObject"
 Set objFSO = CreateObject(fsys)
 Set objTextFile = objFSO.OpenTextFile(alltext,1)
 utc=timestamp : WScript.Echo utc
 Sapi.speak minute(Time)
 c=0

 While not objTextFile.AtEndOfStream
  strNextLine = objTextFile.Readline
  ix=InStr(strNextLine,"#")
  If ix=20 Then
    sa=Split(strNextLine,"#")
  Else
    ix=InStr(strNextLine,"@")
    If ix=20 Then  sa=Split(strNextLine,"@")
  End if
  stamp=left(strNextLine,4) '2158
  If utc=stamp And ix=20 Then
    dbs=Mid(strNextLine,6,3)
    s=Split(sa(1)," ")
    If s(1)="CQ" And s(2)="DX" Then
s(1)=s(1)+s(2):s(2)=s(3)
    If s(1)="CQ" Or s(1)="CQDX" Then sp=s(1)+" from
"+GetCallsign(s(2)) Else sp=GetCallsign(s(2))+" calls
"+GetCallsign(s(1))+" with "+s(3)
    out=dbs+" "+sp
    WScript.Echo out
```

```
     If Abs(cdbl(dbs))>19 Then Sapi.Speak out Else Sa-
pi.Speak sp
      c=c+1
   End If
 Wend
 WScript.Echo c&" calls"&vbCrLf
 If second(Time)<58 Then Sapi.Speak "there were
"+CStr(c)+" calls."
 sec=Second(Time)
 If sec<30 And c > 8 Then sapi.speak "Due to the numerous
decodings from within the last minute and the surprising
detailed data, the current minute decodings are skipped."
 If sec<30 And c > 8 Then sapi.speak " Next decoding is
espected to begin in round about "&CStr(52-Second(Time)-
5)&" seconds."
End sub

Sub showCallsign (a, sapi,sp)
 Set WshShell = WScript.CreateObject("WScript.Shell")
 WshShell.SendKeys "{ESC}"
 WshShell.AppActivate "DX Atlas"
 WScript.Sleep 200
 WshShell.SendKeys "^r"
 WScript.Sleep 200
 WshShell.SendKeys a+"{TAB}{DOWN}{DOWN}{DOWN}{DOWN}^f"
Sapi.Speak sp
 WshShell.SendKeys "{ESC}"
End Sub

' List the UTC Time on a Computer
Function timestamp
 Set objWMIService = GetObject("winmgmts:" _
    & "{impersonationLevel=impersonate}!\\" _
    & ".\root\cimv2")
 Set colItems = objWMIService.ExecQuery("Select * from
Win32_UTCTime")
 For Each objItem in colItems
 h=objItem.Hour: timestamp=CStr(h)
 If h<10 Then timestamp="0"+timestamp
 m=objItem.Minute: If m<10 Then timestamp=timestamp+"0"
 timestamp=timestamp+CStr(m)
 Next
End Function

' List Items in the Local Application Data Folder
Function alltext
```

```
 Set objShell = CreateObject("Shell.Application")
 Set objFolder = objShell.Namespace(&H1c)
 Set objFolderItem = objFolder.Self
 alltext=objFolderItem.Path+"\WSJT-X\ALL.TXT"
End Function


Function GetCallsign(sign)
 Const find="<p class=""haminfoaddress""><b style=""text-
shadow: 0px 1px 0px #f1f1f1, 0px 1px 3px #999; "">"
 GetCallsign=sign
 If sign ="DX" Then Exit Function
  If sign ="RRR" Then Exit Function
   If sign ="GL" Then Exit Function
    If sign ="73" Then Exit Function
     If sign ="RST" Then Exit Function
     If sign ="PSE" Then Exit Function
 jt65=findCallsign(sign)
 If jt65<>sign Then
  GetCallsign=jt65
  Exit Function
 End If

 strURL = "http://qrzcq.com/call/"+sign
 Set objHTTP = CreateObject("WinHttp.WinHttpRequest.5.1")
 objHTTP.Open "GET", strURL:
 On Error Resume next
 objHTTP.Send
 s= objHTTP.ResponseText:  ' WScript.Echo Len(s)
 if objHTTP.Status = 200 Then
   ix=InStr(s,find)
   s=Mid(s,ix+Len(find),100)
   sa=Split(s,"<")
   If UBound(sa)>3 Then
     city=Mid(sa(3),6)
     country=Mid(sa(4),6)
     name=sa(0)
     'GetCallsign = name+" from "+city+", "+country
     writeCallsign sign,name,city,country
     WScript.Echo "New: ---------------------- " +sign+"
- "+name+" ----------------------"
     GetCallsign = name+" from "+""+country
   End if
 End if
 Set objHTTP = Nothing
End Function
```

```
' Verify that a File Exists
function createJT65
 Set objFSO = CreateObject("Scripting.FileSystemObject")
 If objFSO.FileExists(CALLFILE) Then
    Set objTextFile = objFSO.OpenTextFile(CALLFILE, 1)
    While Not objTextFile.AtEndOfStream
     strLine = objtextFile.ReadLine
     If inStr(strLine, "|") Then i=i+1
    wend
    WScript.Echo cstr(i)+" entries found in "+CALLFILE
 Else
    Wscript.Echo "Callsign logfile did not exist, creat-
ed: "&CALLFILE
    Set objFile = objFSO.CreateTextFile(CALLFILE)
 End If
End function

' Writing String Content to End of Existing Text File
Sub writeCallsign(sign,name,city,country)
 Const FOR_APPENDING = 8
 strFileName = CALLFILE
 strContent  = sign+"|"+Name+"|"+City+"|"+Country+vbCrLf
 Set objFS = CreateObject("Scripting.FileSystemObject")
 Set objTS = ob-
jFS.OpenTextFile(strFileName,FOR_APPENDING)
 objTS.Write strContent
End sub

' Read a Comma Separated Values Log
Function findCallsign(sign)
 Const ForReading = 1
 findCallsign=sign
 Set objFSO = CreateObject("Scripting.FileSystemObject")
 Set objTextFile = objFSO.OpenTextFile(CALLFILE,
ForReading)
 Do While objTextFile.AtEndOfStream <> True
  strLine = objtextFile.ReadLine
  If inStr(strLine, "|") Then
    sa = split(strLine, "|")
    If sa(0)=sign Then
     name=sa(1)
     City=sa(2)
     Country=sa(3)
     'findCallsign = name+" from "+city+", "+country
```

```
     If InStr(country,"Netherlands")>0 Then  country="The
"+country
     findCallsign = name+" from "+""+country
     Exit function
   End if
  End If
 Loop
End function
```

The script searches the time-stamp only, not the date, thus All.txt should be deleted once every day to prevent reporting any messages from previous days.

3.8   AERONAUTICAL RADIO TIME ANNOUNCEMENT

3.8.1   SORCERER/REALTERM/RFO/TCP-IP

At amazons preview to [1] can be seen how a time announcement is done using an Android device and *rfo*-Basic. In Central Europe a radio-clock mostly is based on the transmitter DCF77, using special reception modules. In this composition there is a completely different approach to anounce the exact time on an Android smartphone by speech.

3.8.2   TRANSMITTER

Time-source is not the signal from Frankfurt/Main, the in Europe well-known time transmitter on 77.5 kHz, but flight radio signals of HF transmissions on short wave radio. Aircraft and ground stations still transmit worldwide data using the *high-frequency data link*, in short HFDL, with data rates from 300 bps, including exact time data. These sources can be received using an appropriate ssb-radio hardware, but can be heard by anyone via the Internet by WEB-SDR. The Internet, however, is to be understood as a last resort only.

3.8.3   DECODER

Using PC-HFDL or Sorcerer, sophisticated Windows software is available, and both of them support log files - write their output or decodings into a readable text file. Reading from such log files and evaluating data is shown in other sections of this chapter.

A special feature of the Sorcerer software is that it can output the decodings directly and instantly via TCP/IP - a not to be underestimated advantage when it comes to synchronous data.

```
Unable to load grund station names from registry
[HF GROUND STATION CHANGE -> SHANNON - IRELAND]
 15:12:46 UTC SHANNON - IRELAND  DB = 49  SV = 0  GS UP LIGHT  OFFSET 6
SHANNON - IRELAND UTC LOCKED Active freqs        3       5
KRASNOYARSK - RUSSIA UTC LOCKED Active freqs         4
AL MUHARRAQ - BAHRAIN UTC LOCKED Active freqs        2        4
```

Figure 154
Decoding an HFDL time transmission from Ireland at 8942 kHz

3.8.4 CONNECTOR

The decoder provides data via TCP/IP to the local host 127.0.0.1, this is the computer on which the software is running. To direct this data to the local or global network, *RealTerm* is the tool. Without any line of code it is able to redirect a connection from port A to a port B. This even works for serial COM ports, but they are not used here. Thus *RealTerm* can connect to a router using via WLAN/LAN and to an Android device, as an example.

3.8.5 RECEIVER

A Galaxy Note smartphone receives the decoded data of the radio signal via the decoder through TCP/IP as text and there a little *rfo*-Basic program extracts the current and highly accurate time in UTC. It also establishes the TCP/IP connection to the Windows tablet. With a little routine a conversion from UTC to CEST takes place, to prevent alarms go off too early.

At the end of the listing the speech is processed using a little constant time correction, for exact match of the announcement beep as seen on a control-DCF-radio-clock reading.

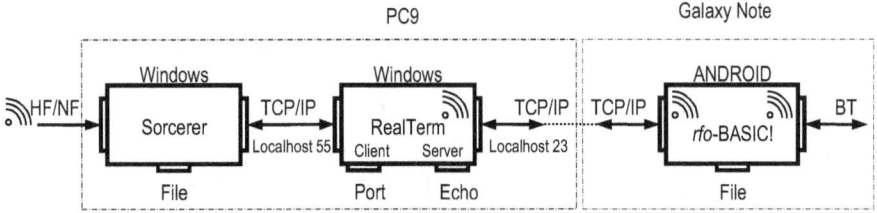

Figure 155: Aeronautical Radio Time Announcement

Necessary for the replica:

- Source: shortwave radio with SSB or Firefox / WebSDR
- Decoder: Sorcerer v1.01 on Windows
- Mediator: RealTerm on Windows
- Receiver: rfo-Basic with specified program

and corresponding Android and Windows devices (here in use: Dell Venue 8 Pro and Samsung GT N7000).

In the summer time, the signal from Shannon/Ireland in Western Europe can be received well in the afternoon at 8942.00 kHz during daytime. If, for example, a Sony ICF2001D is set to 8,942.9 kHz USB, now and then a kind of signal burst can be heard, which is reminiscent of the first data records on music cassettes of home computing - a data packet. A short 1400 kHz tone initiates the mostly just 1.8 seconds short transmission. For WEB-SDR a search finds the wideband SDR website of the University of Twente and there is the setting 8942/USB.

The built-in microphone of the Windows tablet serves as the recording source for the decoder Sorcerer. At *File/Options*, the source can be adjusted, at Add Decoder *ARINC 625* mode from *PSK* has to be selected to see the messages

```
Unable to load system database file
```

```
Unable to load grund station names from registry
```

are not important in this scope. If the microphone is set properly and one of the checkboxes is checked, the first decoding could appear on the screen.

To see the decoded time-data of a ground-station transmission, at least *SPDU* has to be checked. The tab output in Sorcerer offers forwarding and logging/storage of decodings. Selecting *TCP server (Ansi)* opens the specification of the port for data-transfer. Typing a 55 Sorcerer listens to port 55 on the localhost 127.0.0.1 serving the data on demand. *RealTerm* on the same PC configures at the tab port as a source 127.0.0.1:55 because there Sorcerer is listening to provide decodes. Now *RealTerm* does a great job providing an *Echo* (tab) Port as telnet Server, so the incoming data is now accessible from outside the local host to any callers on port 23. This might be a smartphone.

The smartphone as a receiver running an IP-Client may initially check the TCP/IP link by reading the decoded data there. The rest up to the speech then is done by a short *rfo*-Basic program. The time and beep is shown in the lines below

```
TTS.SPEAK  u$
TONE 1000,100
```

Here the entire *rfo*-Basic-Listing

```
TTS.INIT
ps=10
Port$="23"

SOCKET.CLIENT.CONNECT "192.168.178.23",VAL(port$)
SOCKET.CLIENT.STATUS r
IF r THEN
 PRINT "Connected to ";
 SOCKET.CLIENT.SERVER.IP a$
 PRINT a$+" Port "+port$
ELSE
 END
ENDIF

REM SOCKET.CLIENT.WRITE.LINE "."

DO
 DO
  SOCKET.CLIENT.READ.READY flag
  PAUSE ps
  IF CLOCK() > maxclock
   IF ps>10 THEN PRINT "(wait)"
   ps=100
   maxclock = CLOCK() + 5000
  ENDIF
 UNTIL flag
 SOCKET.CLIENT.READ.LINE line$

 IF LEN(line$)>20 THEN
  IF IS_IN(":",line$)=14 THEN
   REM dcf39
   u$=MID$(line$,16,8)
   PRINT u$
   TTS.SPEAK u$
  ELSE
   REM hfdl
   IF MID$(line$,1,1)=" " THEN
    IF MID$(line$,4,1)=":" THEN
     u$=MID$(line$,2,8)
     INCLUDE utc.bas
```

```
   PRINT u$
   TTS.SPEAK  u$
   TONE 1000,100
     ENDIF
   ENDIF
  ENDIF
 ENDIF
 PRINT line$
 ps=10
UNTIL false
```

First speech is initialized, followed by delay-variable and a value of 10. Next is to connect to the Windows tablet using the local IP 192.168.178.23 and port 23. The Dell Venue 8 Pro got this IP from a router. Port 23 (telnet port) connects to *RealRerm*. The main loop ensures that the connection is maintained, incoming lines are scanned for time-stamps (Section HFDL) - and if found announcing the converted time by voice. For reasons of clarity, simplified translation of time takes place in an outsourced routine that is included in the listing. The *utc.bas* file contains the following lines and is stored the *rfo* source directory.

```
REM Start of BASIC! Program UTC TO LOCAL EU
REM U$ in and out
TIME TIME(),y$,mo$,d$,h$,m$,s$,d,dst
t=TIME(y$,mo$,d$,MID$(u$,1,2),MID$(u$,4,2),MID$(u$,
7,2))
REM 8 sec latency
t=t+3600000+dst*3600000+8000
TIME t,y$,mo$,d$,h$,m$,s$,d,dst
u$=h$+":"+m$+":"+s$
```

3.9   DCF39 TIME ANNOUNCEMENT

Parallel to the Central European time transmitter DCF77 serving the well-known RF-clocks, three additional long-wave transmitters, operating ripple control systems, transmit in this frequency range (https://de.wikipedia.org/wiki/DCF39). These sources can be decoded to get the exact time using Sorcerer to supply a smartphone for the purpose of time-announcement. For receiving 77.5 kHz special receivers and antennas are in use, however, this DCF39-signal eventually can be caught by an old long-wave radio, as these devices are waiting for new tasks anyway since the shutdown of many AM stations in Central Europe. The usually at 150 kHz beginning longwave band - even on expensive shortwave receivers - only differs very slightly to the transmission frequency 139 kHz of the DCF39 from Magdeburg. Either you're lucky and the analogue radio without PLL is slightly detuned at the lower end of the long wave range, or the proper coil or the right trimmer-capacitor is to be purposefully slightly adjusted. To get an usable decodable audio signal from a SSB transmission a carrier-frequency is required. Using one of the components in this book this can be done by a trick:

http://shelvin.de/ein-rechtecksignal-mit-hoher-frequenz-mit-dem-arduino-ausgeben

On his webpage Shelvin shares how to get an Arduino to generate this needed carrier-frequency in this range at one of its pins. By connecting this output to a piece of wire creates a small transmitter, whose frequency is mixed with the DCF39 signal of the radio at a suitable distance that a sound sequence is produced, the decoder Sorcerer e.g. can decode by the built-in microphone of the Windows tablet. (Alternatively a SSB-shortwave-radio or WEB-SDR are perhaps an easier source for testing). The appropriate decoder-mode *EFR* is found in Sorcerer in the *FSK-*Section of *Add a decoder.*

```
CURRENT TIME: 20:03:52  Thu 11.08.16  Daylight saving time
CURRENT TIME: 20:04:02  Thu 11.08.16  Daylight saving time
[ C=A0 A=20 CI=00 LEN=16 ] CD CB E3 F6 38 79 69 5B 8C 3C DE
86 C6 A5 B8 AB   "
CURRENT TIME: 20:04:12  Thu 11.08.16  Daylight saving time
CURRENT TIME: 20:04:22  Thu 11.08.16  Daylight saving time
CURRENT TIME: 20:04:32  Thu 11.08.16  Daylight saving time
CURRENT TIME: 20:04:42  Thu 11.08.16  Daylight saving time
CURRENT TIME: 20:05:02  Thu 11.08.16  Daylight saving time
CURRENT TIME: 20:05:12  Thu 11.08.16  Daylight saving time
CURRENT TIME: 20:05:22  Thu 11.08.16  Daylight saving time
```

```
[ C=B0 A=20 CI=00 LEN=16 ] 95 BC 41 4C 9B 5E 84 7A 38 24 B8
62 CE CC 85 7A   "
CURRENT TIME: 20:06:12   Thu 11.08.16   Daylight saving time
CURRENT TIME: 20:06:22   Thu 11.08.16   Daylight saving time
CURRENT TIME: 20:06:32   Thu 11.08.16   Daylight saving time
CURRENT TIME: 20:06:42   Thu 11.08.16   Daylight saving time
CURRENT TIME: 20:06:52   Thu 11.08.16   Daylight saving time
CURRENT TIME: 20:07:22   Thu 11.08.16   Daylight saving time
```

Figure 156: Hex ripple control data and exact time from DCF39

From the point selecting the Sorcerer-TCP/IP port, everything else behaves exactly as described in the *Aeronautical Radio Time Announcement* - one section above -, for *RealTerm* identical applies. Even the *rfo*-Basic listing can perform its duties unchanged as there the DCF39 broadcast signal decoded by Sorcerer has been taken into account.

3.10 MEASUREMENT TELETYPE

3.10.1 PC-HFDL/FILE/NETCAT/TAIL - TCP/IP

A telecommuter/teletyper used to be an electromechanical device using 50 baud to spread news of a news agency. Today such messages are read on the smartphone in a browser.

However, if own measurement data of a technical software is to be transferred, it can be individually configured with the components discussed here. The only condition is, that data is continously written to a textfile aka log-file. This file can be transmitted to a remote device via TCP/IP, to obtain a kind of reading telegraph.

This solution displays live incoming data on a smartphone. The data source is the decoded flight data of a stream received from shortwave HF using the free software PC HFDL. For reconstruction of this interaction, a short-wave receiver is required, but for testing purposes without hardware "WebSDR" internet will do. The connection path is similar to the interaction *Aeronautical Radio Time Announcement*. In difference the decoder software or the measuring program cannot provide its data directly via TCP/IP, but instead constantly a textfile is updated by appending data.

The newly added data in the file will then be sent into the network by means of this book. For this, a script in *VBScript* can be written - as here in the next interaction - or a search in the Internet finds the small tool for the command line: *tail.exe*. This tool is included in the old, but still available Windows Resource Kit (SDK Rootkit Windows 2003 from Microsoft). The 6.5 KB tool outputs the last 10 newly added lines of a log file, as often is used in computing.

The section on *NetCat* describes how to operate the command line on Windows computers operated over TCP/IP. If *tail.exe* is called using appropriate parameters by the command line *NetCat* redirects output from tail to the calling remote device - a smartphone.

3.10.2 IMPLEMETATION

At the beginning there is the software PC-HFDL as a data source. The unregistered version will run for a few minutes and finishes showing a request for registration without requiring user activity. The program generates a log file in textformat *C:\Temp\PCHFDL\logfiles\May03.txt*. The same log-file will be updated even after a (auto-)restart of PCHFDL.

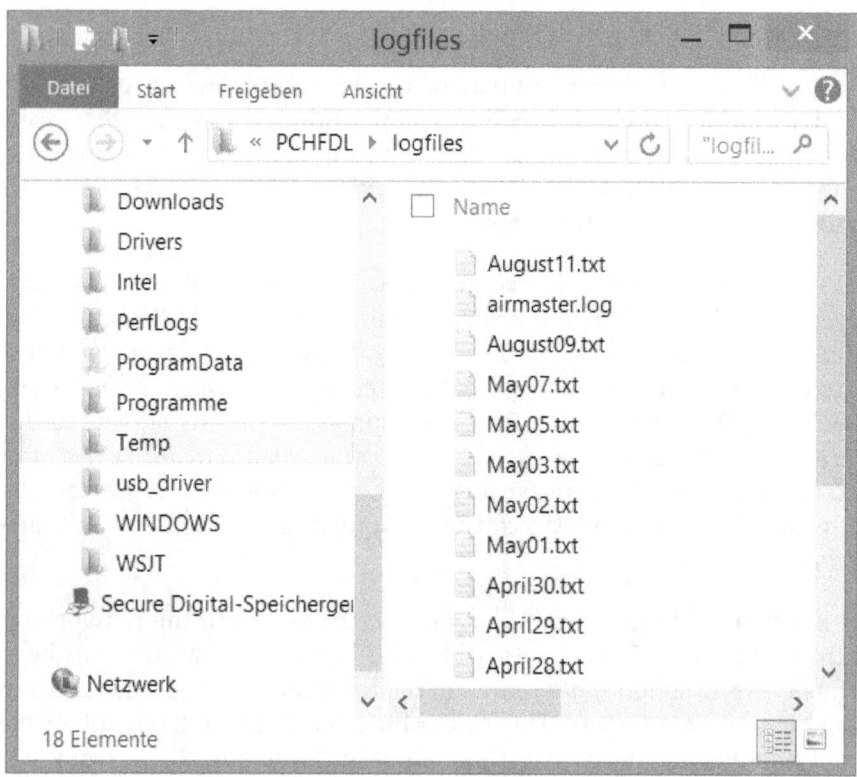

Figure 157:
Content is constantly updated and contains information on air traffic

[HFNPDU FREQUENCY DATA]
19:30:10 UTC Flight ID = SAA585 LAT 29 13 43 S LON 30 49 37 E
GS ID 15 AL MUHARRAQ - BAHRAIN UTC LOCKED
Propagating frequencies   11312 KHz   10075 KHz
Tuned frequencies   21982 KHz   17967 KHz   13354 KHz   11312 KHz   10075 KHz   8885 KHz
GS ID 3 REYKJAVIK - ICELAND UTC LOCKED
Propagating frequencies   11184 KHz
Tuned frequencies   17985 KHz   15025 KHz   11184 KHz   8977 KHz   6712 KHz   5720 KHz   3900 KHz
GS ID 2 MOLOKAI - HAWAII UTC LOCKED
Propagating frequencies

```
Tuned frequencies   21937 KHz   21928 KHz   17934 KHz   17919 KHz   13324 KHz   13312 KHz   13276 KHz   11348 KHz
11312 KHz   10081 KHz   8936 KHz   8912 KHz
GS ID 16 AGANA - GUAM UTC LOCKED
Propagating frequencies
Tuned frequencies   13312 KHz   11306 KHz   11288 KHz   8927 KHz   6652 KHz   5451 KHz
GS ID 4 RIVERHEAD - NEW YORK UTC LOCKED
Propagating frequencies
Tuned frequencies   21931 KHz   17952 KHz   17934 KHz   17919 KHz   13276 KHz   11387 KHz
```

Figure 158: Frequencies and ACARS-Data using HFDL

On the desktop there is a file *ncCmd.bat* containing

```
C:\Temp\nc\nc -L -p 34 -e cmd.exe
```

and run by double-clicking. This line is discussed in the *NetCat* section in detail. Thus, *cmd.exe* is executed as soon as a TCP/IP connection on port 34 is established. Since the controller now is the client, it can start *tail.exe* and thus read the incoming data or log lines and possibly process as desired.

```
C:\Temp\nc\tail -f C:\Temp\PCHFDL\logfiles\May03.txt
```

An example of further processing of such data on the smartphone can be seen in the interplay *Aeronautical Radio Time Announcement* in this chapter just above. Using the same pattern, the log files of the decoder software Sorcerer, WSJT-X, SeaTTY can be transferred.

To meet the headline of this section the decoding software SeaTTY can be used to display the RTTY transmission (radio tele type) of the German Weather Service DWD on short- and longwave (147 or 4582 kHz in Europe) using classic 50 baud speed (50 characters per second) on a smartphone by TCP/IP.

3.11 HFDL-Podcast – Flight Announcement

3.11.1 Sorcerer/VBS/Speech/NetCat-TCP/IP/Cmd.Exe

An acoustic flight-announcement, including flight number, airline, start- and destination airport, is shown in this interaction. The example is processing information from a log file during active data input. The solution presented here is predominantly written in *VBScript* and runs on a Windows tablet. Finally, this information will be displayed on a smartphone via TCP/IP.

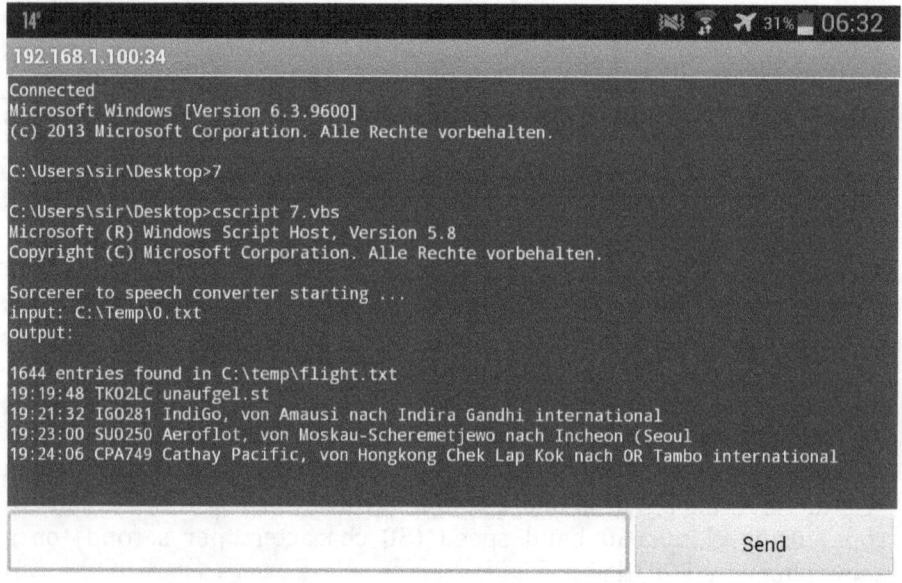

Figure 159: Smartphone displaying flight data while tablet speaks incoming data by command 7

The data source is - as the interaction *Aeronautical Radio Time Announcement* - the free decoder for HFDL signals Sorcerer v1.0.1. In contrast to the local solution, the log file for evaluation approach is drawn here is representative of similar software for measurement tasks. In order to monitor the decoded data of the ARINC 635 module in a log-file, a required file name can be specified by Open Disk Log (ANSI) (disk icon). The following script expects this file as *C:\Temp\0.txt*.

Every 15 seconds the length of this file is checked by the script. Using the variable LastIndex (*li*) the last lines only added are examined whether a flight has been reported. If so, the flight number is evaluatedvand and the belonging information is stored in a local database (similarly, the call sign in interaction JT65 podcast), to let this podcast run after an initial first feeding of the local flight database by Internet data from the source flightaware.com even offline. Internet research and database routines are mostly identical to the versions in conjunction JT65 podcast and are described there in detail.

The constructed message here includes flight number, airline and departure/destination airports, a corresponding adjustment can be done. In the log-file of the decoder a flight looks like this:

```
[MPDU 19:43:58 AIR LOGON SLOT 1 300 BPS ]
[HFNPDU FREQUENCY DATA]
19:23:00  UTC  Flight ID = SU0250  LAT 59 29 46  N  LON
52 56 52  E
```

Based on the time stamp and the fixed expressions ID, LAT and LON the line in question can be identified from the data stream and insulate the flight number SU0250. Lying information on this number is first done locally before an internet search is tried in order to get the missing desired additional data. On failure, the flight is marked as unsolved.

The script first sets some filenames, creates or checks the local database and defines global variables. The number of read lines is stored in *i* (for Index), the previous number of lines read in *li* (LastIndex). The lines are stored in the array *arrFileLines()* after calling *readlog*.

The function *newplane* examines each of the newly added lines matching the expected format of a new flight and reports it. Speech is initialized accordingly. Next the flight-number is isolated and passed to the *Get-Flight*-function, returning the desired information of that line. In case of SU0250, the following line is located in the *flight.txt* file, the local database:

SU0250|Aeroflot (SU) #250 |Moskau-Scheremetjewo (Moscow RU) - SVO / UUEE|Incheon (Seoul (Incheon) KR) - ICN / RKSI

The Aeroflot plane flies accordingly from Moscow to Seoul.

Next is the processing of this data for voice and text output - in English and in German. The statements

```
WScript.Echo utc+ name+" "+gr:   Sapi.Speak en
```

output in the desired text. At the end of the main loop, the script waits using *waitlog* for the next data. The German textoutput in an offline session looks like this:

```
Microsoft (R) Windows Script Host, Version 5.8
Copyright (C) Microsoft Corporation. Alle Rechte vorbehalten.

Sorcerer to speech converter starting ...
input: C:\Temp\0.txt
1644 entries found in C:\temp\flight.txt
19:19:48 TK02LC unaufgelöst
19:21:32 IGO281 IndiGo, von Amausi nach Indira Gandhi international
19:23:00 SU0250 Aeroflot, von Moskau-Scheremetjewo nach Incheon (Seoul
19:24:06 CPA749 Cathay Pacific, von Hongkong Chek Lap Kok nach OR Tambo interna-
tional
19:24:04 SU1406 Aeroflot, von Moskau-Scheremetjewo nach Jekaterinburg
19:23:00 XX0082 unaufgelöst
19:29:34 ONE630 unaufgelöst
19:29:24 SU0512 Aeroflot, von Moskau-Scheremetjewo nach Teheran-Imam Khomeini
19:30:18 SU0509 Aeroflot, von Ben Gurion nach Moskau-Scheremetjewo
19:30:46 KZR918 Air Astana, von Istanbul-Atatürk nach Astana
19:30:52 ONE630 unaufgelöst
19:31:36 TK01DN unaufgelöst
19:32:08 SU2618 Aeroflot, von Moskau-Scheremetjewo nach Brüssel-Zaventem
19:32:18 SU0509 Aeroflot, von Ben Gurion nach Moskau-Scheremetjewo
19:33:12 W30108 unaufgelöst
19:33:28 HVN54 Vietnam Airlines, von London Heathrow nach Hanoi
19:34:00 SU2604 Aeroflot, von Moskau-Scheremetjewo nach Madrid-Barajas
19:34:14 ONE617 unaufgelöst
19:34:18 SU0509 Aeroflot, von Ben Gurion nach Moskau-Scheremetjewo
19:34:16 TK0607 Turkish Airlines, von Istanbul-Atatürk nach Jomo Kenyatta inter-
national
19:34:34 CES788 China Eastern, von Rom-Fiumicino nach Shanghai Pudong interna-
tional
19:34:38 THY3LU THY3LU , von Istanbul-Atatürk nach Helsinki-Vantaa
19:34:48 SU2108 Aeroflot, von Moskau-Scheremetjewo nach Vilnius
19:34:16 TK0607 Turkish Airlines, von Istanbul-Atatürk nach Jomo Kenyatta inter-
national
19:35:26 ONE630 unaufgelöst
19:35:32 SU0504 Aeroflot, von Moskau-Scheremetjewo nach Ben Gurion
19:35:32 D00082 unaufgelöst
19:35:50 LAT unaufgelöst
19:35:52 TK01DN unaufgelöst
19:36:18 SU0509 Aeroflot, von Ben Gurion nach Moskau-Scheremetjewo
19:36:34 MAC472 Air Arabia Maroc, von Venedig-Marco Polo (Venice nach Casablanca
19:36:54 SU2134 Aeroflot, von Moskau-Scheremetjewo nach Istanbul-Atatürk
19:36:56 ME0308 Middle East Airlines - Air Liban, von Beirut nach Kairo
19:37:44 RJA627 unaufgelöst
19:37:46 ONE622 unaufgelöst
19:38:08 ONE627 unaufgelöst
19:38:16 RJA627 unaufgelöst
19:38:18 SU0509 Aeroflot, von Ben Gurion nach Moskau-Scheremetjewo
19:39:04 TK01DN unaufgelöst
19:39:42 CPA289 Cathay Pacific, von Hongkong Chek Lap Kok nach Frankfurt am Main
19:39:56 NMB286 Air Namibia, von Frankfurt am Main nach Intler Flughafen Hosea
Kutako
```

```
19:40:08 TK01DN unaufgelöst
19:40:46 LAT unaufgelöst
19:41:58 PGT3YP PGT3YP , von Antalya nach Istanbul-Sabiha Gökçen
19:40:06 SU1882 Aeroflot, von Moskau-Scheremetjewo nach Manas international
19:25:36 ONE619 unaufgelöst
19:43:52 TK01DN unaufgelöst
19:51:52 W30108 unaufgelöst
19:54:56 KNE057 Flynas, von Riad nach King Abdulaziz international
19:56:50 ETD219 Etihad Airways, von Calicut international (Karipur) (Kozhikode
nach Abu Dhabi
19:57:08 SU2618 Aeroflot, von Moskau-Scheremetjewo nach Brüssel-Zaventem
19:46:12 THY4XN THY4XN , von Prag nach Istanbul-Atatürk
20:00:28 THY4XN THY4XN , von Prag nach Istanbul-Atatürk
20:00:42 KZR918 Air Astana, von Istanbul-Atatürk nach Astana
20:00:56 5K0605 unaufgelöst
20:00:56 SA0204 South African Airways, von John F. Kennedy international nach OR
Tambo international
19:59:50 SA0265 South African Airways, von München Franz Josef Strauß nach OR
Tambo international
20:02:50 AVA011 Avianca, Aerovias Nacionales de Colombia, von Madrid-Barajas nach
Bogotá
20:02:56 CRK949 unaufgelöst
20:02:00 SU1946 Aeroflot, von Moskau-Scheremetjewo nach Almaty
20:03:36 FV6896 Rossiya Airlines, von Sankt Petersburg nach Intler Flughafen
Simferopol
```

Figure 160: Plain text as a script-result

The variable *en* contains the English text for speech output. The overall script looks like this:

```
cl=Chr(13)&Chr(10)
Const FLUGHTFILE="C:\temp\flight.txt"
filename="C:\Temp\0.txt"
interval=15

Dim arrFileLines(),lmod,li

Set fs = CreateObject("Scripting.FileSystemObject")
WScript.Echo "Sorcerer to speech converter starting ..."
WScript.Echo "input: "&filename&cl
createFlightDB

While True
  i = 0
  readlog
  If newplane Then
    Set Sapi = Wscript.CreateObject("SAPI.SpVoice")
    For n = li To i-1 'Beginning = 0
        s = arrFileLines(n):name=Trim(Mid(s,27,7))
        If InStr(s,":")=3 Then
            utc=Left(s,9)
```

```
        flight=GetFlight(name)
        sa=Split(flight,"|")
        u=UBound(sa)
        If u=3 Then
          If InStr(sa(1),"(")Then compa-
ny=Left(sa(1),InStrRev(sa(1),"(")-2) Else company=sa(1)
          If InStr(sa(2),"(")Then
von=Left(sa(2),InStrRev(sa(2),"(")-1) Else von=sa(2)
          If InStr(sa(3),"(")Then
nach=Left(sa(3),InStrRev(sa(3),"(")-1) Else nach=sa(3)
          gr=company+", von "+von+"nach "+nach:
en=company+", from "+von+"to "+nach:

gr=Replace(gr,"Int'l","international"):en=Replace(en,"Int
'l","international"):
        Else
          gr="unaufgelöst":en="unsolved"
        End if
        WScript.Echo utc+ name+" "+gr:   Sapi.Speak en
      End if
  Next
  WScript.Echo "==================================="
  li=i
 End If 'NewPlane
 waitlog
Wend

Sub waitlog
 Do
  WScript.Sleep(1000*interval)
  Loop Until lmod<>lastmod(filename)
  lmod=lastmod(filename)
End Sub

Sub readlog
 Set f = fs.OpenTextFile(filename, 1)
 Do Until f.AtEndOfStream
  Redim Preserve arrFileLines(i)
  arrFileLines(i) = f.ReadLine
  i = i + 1
 Loop
 f.Close
End Sub

Function newplane
 newplane = False
```

```
 For n = li To i-1 'Beginning = 0  last plane li
   s = arrFileLines(n)
   If InStr(s,":")=3 And InStr(s,"LAT")<>0 And In-
Str(s,"LAT 180 0 0  N  LON 180 0 0  E")=0 Then
     newplane=True
     Exit For
   End if
Next
End function

Function lastmod(filespec)
    Dim fso, f, s
    Set fso = CreateObject("Scripting.FileSystemObject")
    Set f = fso.GetFile(filespec)
    s = f.size
    lastmod = s
End Function

Function GetFlight(flight)
 Dim s
 GetFlight=flight
 'WScript.Echo flight
 If flight="LAT" Then Exit function
 find=findFlight(flight)
 If find<>flight Then GetFlight=find:Exit Function

 find="<title>"
 strURL =
"https://de.flightaware.com/live/flight/"+flight
 Set objHTTP = CreateObject( "WinHttp.WinHttpRequest.5.1"
)
 objHTTP.Open "GET", strURL:
 On Error Resume next
 objHTTP.Send
 s= objHTTP.ResponseText:  ' WScript.Echo Len(s)
 if objHTTP.Status = 200 Then
   ix=InStr(1,s,find)
   iy=InStr(ix,s,"</title>")
   a=ix+Len(find)
   b=iy
   If iy>ix And ix>0 Then
       company=Mid(s,a,b-a-14)
            find="<div class=""track-panel-airport"">
<span class=""hint"" title="""
            ix=instr(a,s,find)
            If ix >0 Then
```

```
                        a=ix+Len(find)
                        e=InStr(a,s,"""")
                        von=Mid(s,a,e-a)
                        ix=instr(a,s,find)
                        a=ix+Len(find)
                        e=InStr(a,s,"""")
                        nach=Mid(s,a,e-a)
                        WScript.Echo flight+"--- NEW FLIGHT
FOUND ---"
                        writeFlight flight+"|"+company+
"|"+von+"|"+nach
                GetFlight = flight+"|"+company+
"|"+von+"|"+nach
                End if
        End If
 End if
 Set objHTTP = Nothing
End Function


' Verify that a File Exists
function createFlightDB
 Set objFSO = CreateObject("Scripting.FileSystemObject")
 If objFSO.FileExists(FLUGHTFILE) Then
    Set objTextFile = objFSO.OpenTextFile(FLUGHTFILE, 1)
    While Not objTextFile.AtEndOfStream
     strLine = objtextFile.ReadLine
     If inStr(strLine, "|") Then i=i+1
    wend
    WScript.Echo cstr(i)+" entries found in "+FLUGHTFILE
 Else
    Wscript.Echo "Flight logfile did not exist, created:
"&FLUGHTFILE
    Set objFile = objFSO.CreateTextFile(FLUGHTFILE)
 End If
End Function


' Read a Comma Separated Values Log
Function findFlight(sign)
 Const ForReading = 1
 findFlight=sign
 Set objFSO = CreateObject("Scripting.FileSystemObject")
 Set objTextFile = objFSO.OpenTextFile(FLUGHTFILE,
ForReading)
 Do While objTextFile.AtEndOfStream <> True
  strLine = objtextFile.ReadLine
  If inStr(strLine, "|") Then
```

```
    sa = split(strLine, "|")
    If sa(0)=sign Then
      company=sa(1)
      von=sa(2)
      nach=sa(3)
      findFlight = strLine
      Exit function
    End if
  End If
 Loop
End function

' Writing String Content to End of Existing Text File
Sub writeFlight(data)
 Const FOR_APPENDING = 8
 strFileName = FLUGHTFILE
 strContent  = data+vbCrLf
 Set objFS = CreateObject("Scripting.FileSystemObject")
 Set objTS = ob-
jFS.OpenTextFile(strFileName,FOR_APPENDING)
 objTS.Write strContent
End sub
```

As the text-output can be redirected using *NetCat* to a TCP/IP port, this information can be shown on a smartphone using a TCP/IP client. The speech is still done on the Windows tablet. To get this going the same steps have to be taken as in section measurement teletyper. Calling *NetCat* using

```
C:\Temp\nc\nc -L -p 34 -e cmd.exe
```

executes the command line processor on connect. A smartphone or tablet connects to port 34 via TCP/IP to control windows by command-line. Rather than *tail.exe*, the above script is called from a batch file to redirect output using *cscript* to the standard (console) output. If called directly, the script will display a messagebox on a windows PC for every output, generating issues.

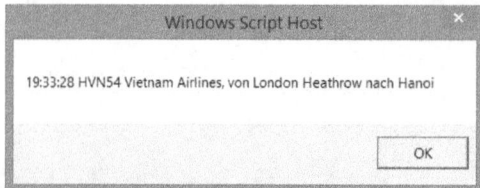

Figure 161: Script message-box is not recommended in this context

If this script lives as *7.vbs* on the desktop and batch-file *7.bat* containing the text *cscript 7.vbs* lives there too, an indirect start of the script is done by typing a '7' followed by enter at the windows-prompt of an TCP/IP-client on a smartphone to start the output, as shown in the figure at the beginning of this section.

3.12 PLANE-HOPPING - RADIO CONTROLLED GOOGLE-EARTH

3.12.1 SORCERER/FILE/VBS/GOOGLE-EARTH

Google Earth can be used offline, if there is sufficient gathered material in the cache from a previous internet session. On this basis, it should be possible, even offline, to view the flight positions received via short-wave data using Google Earth in real time. This built-in DDE feature of the decoder PC HFDL breaks in the unregistered version this interface connection by the automatic termination and after a restart everything has to be reconfigured.

Figure 162: Google Earth and the XML HFDL data

The interaction of the free HFDL decoder Sorcerer and Google Earth using *VBScript* is since mid-2016 published at

http://www.hjberndt.de/dvb/hfdl.html

and is listed here as an another example of how data from updating log files can be evaluated and displayed by *VBScript*.

A local shortwave receiver (or WebSDR) as a data source like in the previous section HFDL-PODCAST is in use. All settings remain the same for Sorcerer, so the log file is written in the temp-directory. The script does similar queries, except that the result is a KML-file, a Google-Earth readable format, indicating the flight positions. Sorcerer 1.0.1 decodes the HFDL signal and puts the result on screen and to a log-file. A script converts the flight positions into a Google Earth readable format. To get the coordinates of a flight, *HFNPDU* has to be checked. The line

18:54:30 UTC Flight ID = BER537 LAT 39 32 49 N LON 2 44 19 E

is reporting the position of flight BER537 with specified universal time. This data has to be converted by *VBScript* to a Google Earth KML format. The script monitors the length of Sorcerers log-file. On change the appended lines are searched for lines starting by a time-stamp, in order to extract the position data and generate a readable Google Earth KML file. On a new flight this file is "executed". If KML-files are linked to Google Earth as the default application, then Earth is started loading this file, or by processing the KML-file the new position(s) just are displayed. This processes in a 15 seconds interval.

```
16:26:26  UTC  Flight ID = MAC341  LAT 41 33 22  N  LON 1 52 55
16:04:06  UTC  Flight ID = SU1404  LAT 180 0 0   N  LON 180 0 0
16:14:42  UTC  Flight ID = SAS773  LAT 52 52 34  N  LON 6 32 57
16:27:12  UTC  Flight ID = SAS773  LAT 54 3 8    N  LON 8 28 37   E
16:28:02  UTC  Flight ID = CCA939  LAT 44 0 39   N  LON 13 8 32
16:28:46  UTC  Flight ID = OMA816  LAT 18 20 10  N  LON 81 10 11
16:29:12  UTC  Flight ID = CSC856  LAT 30 41 9   N  LON 105 8 22
```

Figure 163: Flight scoreboard as console-output

In summary the following steps have to be taken for using the script listed below in this context.

- Run Sorcerer 1.0.1 (allow to defender, else a huge delay occurs)
- Set the SSB receiver to 8942/10081 kHz USB (Shannon)
- Activate the microphone on the tablet and observe whether the signals are audible.
- Select PSK/ARINC 635 in Sorcerer. Checkbox *HFNPDU* checked.
- Select *Open disk log (ANSI)* from the output tab and create a log file with the name *0.TXT* in the directory *C:\TEMP*.

On OK Sorcerer logs all displayed decodings to that file. Now the script can be executed from any folder using the hardcoded path *C:\TEMP\0.TXT* for the input file. Suppose the script is called *6.vbs* and is located on the desktop, this script should be executed indirectly to supress many messageboxes. For this reason a text file named *6.bat* and the text content *cscript 6.vbs* can be created on the desktop. Running *6.bat* generating a kml file *test0.kml* is started and launched using Google Earth.

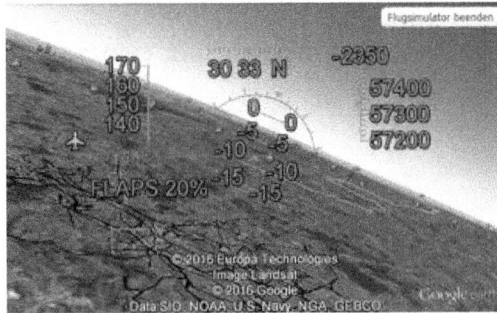

Figure 164: Google-Earth flight simulator

Whenever a new flight is detected, Google Earth zooms in - if only one machine came along - deep at the location of the position data. By watching the incoming positions, most unknown locations on this planet get visible. Even the built in flight-simulator may be activated to fly from the position to other destinations, but beware: offline, a visual flight might be quite difficult, due to the lack of cached data resulting in a dizzy view.

A rough structure of the script could look like this:

- repeat
    - read log-file completely (*readlog*)
    - if new flight data (*newplane*)
        - create XML file header
        - insert position as XML data
        - create XML footer
        - output new flight (*Wscript.Echo*)
        - execute Google Earth using KML-file
    - wait for new log entry (*waitlog*)
- forever

3.12.2 FUNCTIONS AND SUBS

A modified filelength indicates a contentchange of the log file. The *last-mod* function returns the file size.

```
Function lastmod(filespec)
  Dim fso, f
  Set fso = CreateObject("Scripting.FileSystemObject")
  Set f = fso.GetFile(filespec)
  lastmod = f.size
End Function
```

The routine *waitlog* regularly reviews the filelength and returns if a change occurs.

```
Sub waitlog
  Do
    WScript.Sleep(1000*interval)
  Loop Until lmod<>lastmod(filename)
  lmod=lastmod(filename)
End Sub
```

As shown in the above structure the entire LOG-file is read at the beginning. In initial versions of this script the order was slightly different. All flights are investigated next. File reading and memory management are taken from the *VBSEdit* example *Read a text file from the bottom up*.

```
Sub readlog
  Set f = fs.OpenTextFile(filename, 1)
  Do Until f.AtEndOfStream
    Redim Preserve arrFileLines(i)
    arrFileLines(i) = f.ReadLine
    i = i + 1
  Loop
  f.Close
End Sub
```

To test the log file on new planes, it is loaded into memory completely and then checked using the function *newplane*.

```
Function newplane
  newplane = False
```

```
For n = li To i-1 'Beginning = 0   last plane li
  s = arrFileLines(n)
  If InStr(s,":")=3 And InStr(s,"LAT")<>0 And _
  InStr(s,"LAT 180 0 0  N  LON 180 0 0  E")=0 Then
    newplane=True
    Exit For
  End if
Next
End function
```

All of the three global variables are invoked here. The logfile is stored in an array *arrFileLines*, with *li* holding the "last" or previous line number read, and *i* the current number of lines. If the third character is a ":" and "LAT" occurs, it is likely this line holds flight position data. Because some machines show up without coordinates, they are discarded. On a valid new flight the function returns true.

```
Sub Google
  Set WshShell = WScript.CreateObject("WScript.Shell")
  Return = WshShell.Run(kml)
End Sub
```

Google-Earth is run indirectly by calling the newly created KML-file in this Google-routine. KML files have to be assigned to Google-Earth.

The last function *latlon* is a reasonably reliable scan for coordinates in the flight data line for output to the XML file in the main loop.

```
Function latlon(lonlat,s)
    sp0=InStr(s,lonlat)+4
    sp1=InStr(sp0,s," ")       'LAT 50 38 27  N
    lat1=Trim(Mid(s,sp0,sp1-sp0))   '50
    sp2=InStr(sp1+1,s," ")
    lat2=Trim(Mid(s,sp1+1,sp2-sp1))'38
    sp3=InStr(sp2+1,s," ")
    lat3=Trim(Mid(s,sp2+1,sp3-sp2))'27
    lat4=Trim(Mid(s,sp3+2,1))        'N
    latf=CDbl(lat1)+CDbl(lat2)/60.0+CDbl(lat3)/3600.0
    If lat4="S" Or lat4="W" Then latf=-latf
    lat=FormatNumber(latf,6)
    lat=Replace(lat,",",".")
    latlon=lat
End Function
```

The entire log line is passed including degrees, minutes and seconds and *lonlat* whether "LON" or "LAT" is to be converted. The result is a signed decimal as expected by Google-Earth.

3.12.3 MAIN LOOP

The script starts with the declarations followed by the main loop. Including the above routines, the overall script can be created. Assuming this is stored as *6.vbs*, then a corresponding *6.bat* with the line *cscript 6.vbs* is to be created. Again: Executing *6.vbs* without console redirection will result in a lot of msg-boxes, depending on how many flights are found in the Sorcerer log file *0.txt*. The missing part of *6.vbs* looks like this:

```
Const cl=Chr(13)&Chr(10)

'Set Args = WScript.Arguments
'If args.Count<>2 Then
   filename="C:\temp\0.txt"
   kml="c:\temp\test0.kml"
   interval=15
'End if

'filename=Args(0)
'kml=Args(1)
'interval=Args(2)

'-------------------------------------
Dim arrFileLines(),lmod,li

Set fs = CreateObject("Scripting.FileSystemObject")
WScript.Echo "Sorcerer to Earth converter starting ..."
WScript.Echo "input: "&filename
WScript.Echo "output:"&kml&cl

While True
  i = 0
  'waitlog   ' wait for new filelength of logfile
  readlog
  If newplane Then
   'Create XML Document header
   Set fx = fs.CreateTextFile(kml,True)
   x="<?xml version=""1.0"" encoding=""UTF-8""?>"& cl & _
```

```
    "<kml xmlns=""http://earth.google.com/_
     kml/2.0"">"& cl &_
    "<Document>"& cl & _
    "<Style id=""bluepin"">"& cl & _
    "        <IconStyle>"& cl & _
    "          <Icon>"& cl & _
    "
<href>http://maps.google.com/mapfiles/kml/_
shapes/airports.png</href>"& cl & _
    "           </Icon>"& cl & _
    "          </IconStyle>"& cl & _
    "      </Style>"& cl
  fx.WriteLine x
  'Add new placemarks in xml
  For n = li To i-1 'Beginning = 0  one plane li
   s = arrFileLines(n)
   If InStr(s,":")=3 And InStr(s,"LAT")<>0 And _
   InStr(s,"LAT 180 0 0  N  LON 180 0 0  E")=0 Then
    x="<Placemark>"&cl& _
      "<styleUrl>#bluepin</styleUrl>"&cl& _
       "<name>" & Trim(Mid(s,27,7)) & "</name>" &cl& _
       "<description>"&cl& _
       "TIME:  " & Left(s,13) & "<br />" &cl& _
       "</description>"&cl& _
       "<Point>"&cl& _
       "<coordinates>" & latlon("LON",s) & _
   ","&latlon("LAT",s)&",10000</coordinates>"&cl& _
       "<TimeStamp><when>2016-1-1T" & Left(s,8) & _
   " Z</when></TimeStamp></Point>"&cl& _
       "</Placemark>"&cl
          fx.WriteLine x
          'WScript.Echo s
    End if
  Next

  'Write the xml footer
  x="</Document>"&cl&"</kml>"&cl
  fx.WriteLine x
  fx.Close 'XML

  'New Planes with zero LAT LON
  For n = li To i-1 'Beginning = 0
      s = arrFileLines(n)
       If InStr(s,":")=3 Then       WScript.Echo s
  Next
```

```
  li=i
  Google
 End If 'NewPlane
 waitlog
Wend
```

Due to XML strings, the listing looks somewhat torn.

3.13 RHINETOWERCLOCK HF-CONTROLLED

3.13.1 DIGISPARK/HC06/RFO/REALTERM/SORCERER

An Arduino synchronizing the Rhine Tower Clock using a DCF77 RF-module can be seen since long at

http://hjberndt.de/soft/rtctftdcf.html.

This composition continues the Hardware-Section Digispark - Rhine Tower using 50 LED. The aim is to synchronize to a short- or long wave transmitted time source, as the *Aeronautical Radio Time Announcement* HFDL or radio time announcement DCF39. The corresponding expanded schematic looks like this:

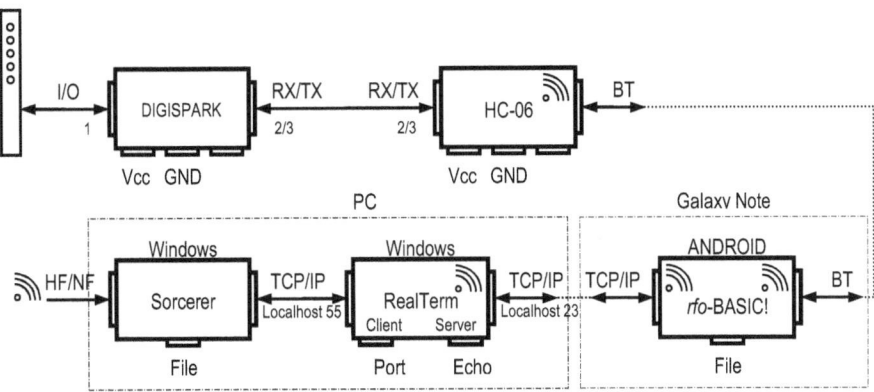

Figure 165: All of the components and their interaction

The RF-signal is demodulated in a suitable radio receiver and is available as audio-source on a Windows tablet. This signal is decoded by the free software Sorcerer and passed as a result directly to a TCP/IP-port on the localhost to the terminal software *RealTerm* on the same PC. This client locally connects using this port to Sorcerer and redirects the connection using its echo function as a telnet server. An Android smartphone or tablet telnets to this "PC9" in the local network using a *rfo*-BASIC program and redirects the time data via Bluetooth to the Digispark using the *HC06* module.

Main actor is the *rfo*-Basic program, establishing the two connections and links them, but in addition also extracts and transforms the time information from the data stream to the required format. Extending prior listings can lead to a new listing looking like this:

```
TTS.INIT
Port$="23"
IP$= "PC9"
include btopen.bas
include ipopen.bas
ps=10

DO
 DO
  SOCKET.CLIENT.READ.READY flag
  PAUSE ps
  IF CLOCK() > maxclock
   IF ps>10 THEN PRINT "(wait)"
   ps=100
   maxclock = CLOCK() + 5000
  ENDIF
 UNTIL flag
 SOCKET.CLIENT.READ.LINE line$
 IF LEN(line$)>20 THEN
  IF IS_IN(":",line$)=14 THEN
   REM dcf39
   u$=MID$(line$,16,8)
   PRINT u$
   BT.WRITE u$+CHR$(13)
   TTS.SPEAK u$
  ELSE
   REM hfdl
   IF MID$(line$,1,1)=" " THEN
    IF MID$(line$,4,1)=":" THEN
     u$=MID$(line$,2,8)
     INCLUDE utc.bas
     PRINT u$
     TTS.SPEAK  u$
     TONE 1000,100
     BT.WRITE u$+CHR$(13)
    ENDIF
   ENDIF
  ENDIF
 ENDIF
 PRINT line$
```

```
 ps=10
UNTIL false
```

For readability reasons the opening of the two connections is included from the external files *BTopen* and *IPopen* located in the same directory. The details are as follows:

```
REM Start of BASIC! Program
bt.open
bt.connect
do
 pause 1000
 bt.status x
 if x= 1 then print "listening ..."
 if x= 2 then print "connecting ..."
 if x= 3 then print "conected to ";
 t++
 if t>20 then
 print "timeout."
 end
 endif
until x=3
BT.DEVICE.NAME dev$
print dev$
```

BTopen.bas include file

```
REM Start of BASIC! Program
PRINT "Connecting to "+ip$
SOCKET.CLIENT.CONNECT IP$,VAL(port$)
SOCKET.CLIENT.STATUS r
IF r THEN
 PRINT "Connected to ";
 SOCKET.CLIENT.SERVER.IP a$
 PRINT a$+" Port "+port$
ELSE
 END
ENDIF
```

IPopen.bas include file

The listing takes into account both the formats for HFDL and DCF39, as supplied by Sorcerer and returns the exact time of short- or long-wave radio over Wi-Fi radio and Bluetooth radio to the RF controlled Radio Tower Clock; pretty radioactive.

BIBLIOGRAPHY

[1] Messen mit dem Smartphone, H.-J. Berndt, Eigene Programme auf Android Tablet und Phone, kindle-ebook ASIN B00CO5TGEK, May 2013

[2] Messen und Steuern mit dem Smartphone, H.-J. Berndt, Bluetooth, USB, RS232, Arduino mit Android Tablet/Phone, kindle-ebook ASIN B00SM1UMQG, January 2015

[3] Messen, Steuern und Regeln mit Word und Excel, H.-J. Berndt / B. Kainka, VBA-Makros für die serielle Schnittstelle, 3., aktualisierte Auflage, Franzis-Verlag GmbH, 85586 Poing, 2001 ISBN 3-7723-4094-6

FIGURES DIRECTORY

GLOSSARY

The Patient's Handbook of Pain Management

Pain Occurs Naturally; Suffering does not!

William E. Ackerman III, MD

Pain Medicine Consultants
Epmo.info
Expert Pain Medicine Opinions

The Patient's Handbook of Pain Management

Copyright © 2008 by William E. Ackerman III, MD

This book is dedicated to my loving wife Carrie, my family, and my staff and to my patients whose frequent questions inspired the content for this book.

Acknowledgments

I wish to acknowledge my patients. Their questions that they asked concerning their pain prompted the writing of this book. If a patient has a basic understanding of their pain pathology, it helps he or she cope better with the pain and helps the treating physician better understand their suffering and ultimately provide better care.

Thank You.

Foreword

Pain management is becoming an important and growing medical specialty.   There is an attitude among individuals suffering from chronic pain that they are no longer willing to suffer pain in silence. Dramatic changes have been made with respect to the understanding of the anatomy any physiology of many painful entities over the past two decades.   New drugs and other modalities are being introduced with increasing frequency.   A pain patient needs to partner with his or her doctor.   The reason for this book is to give a pain patient the basic information necessary to rationally discuss his or her pain with the treating physician. Many primary care physicians have no or minimal pain training.   This pocket-sized book will enable a patient to derive basic pain management knowledge that may be helpful when he or she communicates with their pain management physician.

Table of Contents

1. Overview of Pain Management

Almost everyone experiences pain at some time. Pain can be a natural response to injury and disease. With the advent of pain medicine as a medical specialty, patients no longer need to suffer. Suffering is how our lives are affected. Patients who suffer have significant reductions in the normal joys of their lives. They can not enjoy their families or enjoy recreational activities etc. Their pain affects them emotionally.

Animal models have enhanced our understanding of pain mechanisms and make forward-looking statements as to our proximity to the development of effective mechanism-based treatments. Animal pain models have failed in some aspects as animal models can not always be extrapolated to humans.

Pain is an unpleasant sensory and emotional experience following tissue injury. Your pain management can be expensive as well as ineffective if you do not communicate truthfully with you doctor. Different specialties in medicine practice pain medicine. Chiropractors practice pain management as well. Anesthesiologists manage your pain with injections. Physiatrists manage pain with heat, cold, etc. and do needle studies on you to see if you have nerve damage.

Orthopedic and neurosurgeons can perform surgery on you. A neurosurgeon can place a spinal morphine pump that directs morphine into your spinal fluid. All of these specialists can prescribe drugs to you for pain management. A multidisciplinary pain clinic will have physicians, physical therapists, and psychologists who can collectively treat your pain. The effectiveness of psychosocial interventions for back pain in primary care has been established.

Pain is a complex, idiosyncratic experience. When pain is the primary complaint for seeking medical attention, an understanding of multiple factors is essential in guiding successful treatment. Behavioral medicine, a branch of psychology, has been an integral part of interdisciplinary/multidisciplinary care of pain patients.

Prominent and distressing emotions, cognitions, and behaviors frequently accompany chronic pain. In many cases, these psychological symptoms will be sufficiently severe to qualify the patient for a diagnosis of a mental disorder.

For thousands of years, doctors have been helping to relieve their patients' pain with a variety of medications and treatments. Like other areas of medicine, a new subset of doctors has become specialists in treating pain. The question that you should ask yourself is if any of the modalities such as heat, cold, injections, drugs etc. will actually stop your pain. The answer is no if your pain is chronic. Chronic pain is a disease. Your doctor will strive to provide you with a quality of life. If your pain is acute such as post surgical pain, or after a fall on your hip you should expect significant pain relief. Chronic pain is that pain that persists after your body has healed. Nothing unfortunately will completely eliminate your pain.

The goal of pain management is to decrease your pain so that you can maintain your normal activities of daily living. This means that your pain should not interfere with your work, family or recreation. As a result, if you have chronic pain your goal and your doctor's goal should be to decrease your pain to a tolerable level. Pain management is expensive. Because nothing will completely stop your chronic pain, you will need to follow up frequently with your health care provider. Different treatments will be tried until you begin to have a reduction in your pain.

When you see a specialist or go to a hospital or surgery center, if you have insurance, you will have to pay a co-payment to have a procedure or an examination performed. Because many pain treatments will not benefit you, it is necessary for you to become an informed consumer. You can do this by trying to understand what is causing your pain and what alternative modalities (such as chiropractic, herbs etc.) are available to you. If you complain of pain to your primary care physician, your doctor may refer you to someone who only treats pain. This individual may have no or minimal training or may have extensive training.

Remember that these treating individuals have expenses such as MRI machines, X ray machines to do steroid injections on you, physical therapy equipment etc. This equipment must be paid for. You, the patient will ultimately pay for the equipment. You must therefore, become knowledgeable as to what treatment is necessary to treat your painful condition so that you will not receive treatments that you do not need. This is why communication between you and your doctor is important.

You must be aware that pain management can be an environment where essentially a practitioner in some situations needs only minimal credentials such as a medical license to do potentially harmful procedures to unsuspecting individuals suffering from excruciating pain. This can be a world where most procedures can be done in out- patient surgery centers to avoid the peer review scrutiny of a hospital medical staff. You must therefore, inquire if your practitioner is trained and certified in pain medicine. Approximately forty-eight million Americans suffer from chronic pain.

Americans spend over one hundred billion dollars annually on pain care. One-third of all adult Americans suffer from chronic pain. Over the counter annual analgesic costs amount to three billion dollars. Your pain treatment can bankrupt you. You must be able to identify the ethical pain treatment physicians and clinics as well as the "get rich" schemers. The cost of medical care is rapidly escalating. Employers may not be able to afford health insurance for their employees. The cost of pain management is contributing to the increase in health care costs and amounts to hundreds of thousands of dollars. As a result, there is considerable profit to be made by unethical health care providers that include hospitals as well as physicians.

Unfortunately, along with many reputable excellent pain clinics many pain management centers have sprouted up like weeds throughout the United States staffed by individuals with little or no formal training. One may look in the Yellow Pages of any telephone book in most communities to find an establishment that will manage chronic pain syndromes including cancer pain. Unfortunately, there are no state regulatory bodies, which govern the way these clinics or physicians practice pain management. Patient influence and persuasion to go to a specific pain clinic is noted in many pain treatment center advertisements, which become unethical if anything is done to interfere with patient free choice through intentional deception or distortion. These abuses are becoming sufficiently common and flagrant.

Physicians who themselves do not practice pain medicine such as orthopedic surgeons might refer patients to medical centers that they own to have MRI's and multiple injections done by pain specialists to be performed in an operating room setting. The reader must realize that the facility owner must pay for the MRI scanner, X ray machine, the building and the personnel that it takes to staff the facility. This cost can be millions of dollars annually. However, physicians with shares in these facilities realize handsome profits from ownership in these facilities.

Patients who are unable to pay are frequently excluded from treatment. If the treatment is as good as advertised in the newspaper, on television or on the radio, why are some patients denied care? The sad fact is that this behavior is legal. Consumer protection legislation and patient education will play an increasingly significant role when one decides to choose a pain treatment center and/or physician.

Many pain clinics hire marketing firms to promote their services. Very few advertisements that this author has reviewed in ten large cities, mention the credentials of their physicians. Most advertisements mention only the treatment of many painful entities. These advertisements fail to mention that they will only do invasive care in a facility that they own to enjoy a bigger payday.

You as a patient and consumer should have the choice of the facility where your procedure will be done. In fact, you should insist on it. You need to be aware that many hospital administrators and physicians in the 1990's realized after considerable market research that they could enjoy large profits from pain management. Hospitals could charge fees by having nerve injections performed in the hospital instead of in a physician's office.

Health care costs due to chronic pain are particular high during the first year after the onset of your pain, and remain high compared with health care costs before your pain onset. The majority of chronic pain patients incur the costs of alternative treatments. Chronic pain causes production losses at work, as well as impairment of your non-work activities.

Chronic pain patients must realize that the majority of pain procedures can be done in the physician's office. You or your insurance company will save approximately $1200.00=$1500.00 per procedure. Medicare realizes this and encourages office based pain management. Private insurance carriers now realize that they are literally being fleeced by some pain block facilities.

Medical equipment companies in the 1990's also saw the potential to make considerable sums of money by selling devices that could burn nerves, freeze nerves, place salt solutions on nerves or "melt" disc structures in the back. Unfortunately many of theses devices have no scientific merit. Medical instrument companies also manufactured electric catheters to be placed in the backs of patients that were intended to diminish pain. There may be medical evidence that these devices diminish pain in a number of select patients with specific pathological entities.

Three companies at the time of the writing of this book sell a light scope that is placed into your spine from above the rectum. Allegedly, a physician can identify where one should place a steroid. Unfortunately, some patients have been blinded by this procedure. One company did manufacture a pump device, which does deliver morphine or morphine like drugs into the spinal fluid of cancer patients.

This device did provide significant pain relief and did increase the quality of life in many cancer patients. This company then used their technology to provide pain relief in noncancerous pain patients. Unfortunately, they and the other companies required no training by physicians for the implantation of any of these devices. Outside of the medical field, this activity would attract widespread media attention.

Most medical specialties other than pain management require extra training (usually a minimum of 12 months) in a subspecialty of medicine before a physician may refer to them selves as a specialist. However, in pain medicine if an individual is not trained in a procedure the sales representative will come into the operating room (usually without a patient's consent) and instruct a physician on how to do a potentially dangerous procedure or a physician may go to a weekend course sponsored by the manufacturer and practice on a cadaver and

subsequently be "certified" as an expert in the performance of a procedure.

You should be aware that pharmaceutical representatives may have
access to the prescribing habits of physicians. Physicians are frequently enticed into prescribing a certain medication with luncheons, dinners
and lavish trips provided by a pharmaceutical company. This practice
has been recently documented on a television news story. You should
be aware that many of these medications are not safer, better or cheaper
than existing medicines.

You should also be aware that the American Board of Medical Specialties has a list of specialties recognized as true medical specialties or
subspecialties in the United States. Pain management itself is not
included in this list. However, the American Board of Anesthesiology
added qualification in pain medicine is recognized. Physicians other
than anesthesiologists such as physical medicine and rehabilitation,
neurologists etc. can become board certified. This should tell you
"Buyer beware!" Many patients use the Internet to access information
regarding pain management. In my practice, less than ten percent of
patients obtain medical information from an academic source such as
the National Library of Congress from computer sources.

The purpose of this book is to inform you that outstanding pain centers
do exist throughout the United States but that the "buyer must beware"
men tality must be considered with respect to some "injection mills"
staffed by some profit only driven individuals with minimal training.
The reader must furthermore, be aware that a small number of pain
clinics will prescribe potent narcotic medications until an individual
becomes addicted. This behavior is not unlike the drug dealer on the
street.

At the time of your addiction, you must agree to have expensive
injection procedures that must be done biweekly with the threat that
they are noncompliant if they refuse injections and will have their
narcotic prescriptions discontinued. This practice is called "hookem,
stickem, and bootem".

The street drug hustler is incarcerated when caught. Nothing happens
to the physician who intentionally causes a patient to become addicted
to narcotics.   On occasion a state medical board will investigate

unscrupulous practices. An unscrupulous physician can legally get by with causing a patient's death or in many instances mutilation. The reader of this book will be aware that there is no rationale to ever having any "series" of injections done unless benefit is noted with each injection.

Individuals must be aware that some physicians on occasion will entice a patient to have more than one procedure even if the first procedure was ineffective. Usually the physician will recommend three procedures. If the first procedure was properly done why have two more procedures? Two more procedures will make the physician's house payment but probably will not have any affect on your chronic pain.

Small pain centers, which have few patients, will entice patients to have three or more injections in a weekly or biweekly series only to increase facility and physician revenues. You will need to learn to distinguish the physician who practices science-based medicine from the physician who is a charlatan.

After completing this book, you should be able to discern what modalities can benefit you and which modalities benefit only the physician. This book may save the person who is suffering from chronic pain, significant time and money by learning what modalities can actually help an individual suffering from chronic pain. If a nerve block is done in the physician's office there is no facility charge. This will save you a facility co-payment.

This book will enable you to gain a basic knowledge of the pathophysiology (the cause of your pain) and treatment of various chronic pain entities. This will enable you to become a team member with your doctor. This knowledge should help prevent you from becoming a potential victim by avoiding the incompetence of certain physicians who claim to be "pain medicine specialists" and by avoiding procedures that are possibly dangerous or have absolutely no scientific merit.

For those individuals who do not wish or have the time to read this book or a similar book you may be leaving your overall well being to the "luck of the draw". You must be aware that many referring physicians have minimal knowledge as to what constitutes ethical pain medicine as pain management is not taught in most medical schools. This book is written for the layperson suffering from chronic pain. It is hoped that you will derive an understanding of the various painful conditions that may affect you or a family member and to relate to you what potentially effective treatments are available.

2. Pain Anatomy and Physiology

The word "pain" is derived from the Latin word poena that means punishment. St. Augustine wrote in the 5th century that all diseases afflicting Christians were derived from demons. Ancient tribal concepts of pain were based on beliefs that evil spirits were sent as punishment from their gods to invade one's body and cause severe pain. In the book of Genesis, Eve was condemned to pain during childbirth as a result of her encounter with the devil in the Garden of Eden. It has been reported that a shaman could suck an evil spirit from a wound to decrease one's pain.

The ancient Greeks such as Aristotle were the first individuals who believed that pain was derived from various nerves in the body. The exact cause of pain was unknown to them. Unfortunately, not unlike ancient times, the diagnosis and treatment of many chronic painful conditions today remains mostly guesswork. Pain medicine is for the most part subjectively based, because pain is a subjective symptom while other medical specialties are based upon objective medical evidence. Pain in general is not bad. Pain is a protective mechanism that warns you that your body has something wrong at some location. The sensation of pain tells you to stop activity or to at least slow down your activity. For example if you sprain your ankle, your pain is a warning for you not to put weight on that leg. The International Association for the Study of Pain defines pain as" an unpleasant sensory and emotional experience associated with tissue injury as a result of trauma (eg. bone fracture) or disease (e.g. cancer, shingles).

Pain has psychological effects in some instances especially when pain is severe. Pain may cause anxiety and depression. Acute pain is associated with injury, bone fractures, surgery or sprains and strains. Once these entities have healed sometimes, the pain continues. Arthritis is another example of chronic pain. Arthritic pain is caused by continuous joint destruction. However, once the pain becomes chronic, your pain it becomes a problem. Not only does pain become a personal problem but pain can become a social problem with creation of family problems, loss of self esteem and lost wages. Fibromyalgia patients have alterations in CNS anatomy, physiology, and chemistry that potentially contribute to the symptoms experienced by these patients

The purpose of this chapter is to present you with the basic anatomy and physiology of painful sensations. Pain impulses are in essence, electrical signals that travel from various areas of your body such as the extremities, heart, appendix etc. to the spinal cord and eventually reach the brain where the pain signals are processed like data in a computer. The brain is like a computer hard drive, which stores painful experiences that ultimately results in the suffering associated with chronic pain. Pain is produced by unpleasant stimuli to nerve endings throughout the body which include chemical, extreme heat cold and mechanical injury. These nerve endings are silent until mechanical, heat or cold injures tissue. In order to experience pain we need these pain receptors and the nerve fibers that transmit pain to the spinal cord and then to the brain.

Nerves, which conduct pain impulses to the spinal cord, are composed of neurons (nerve cells) that make up nerve fibers that form neurons. Two common pain fibers are the C fibers and the A-delta fibers. A-delta fibers conduct fast onset sharp pain impulses. The C fibers conduct slow onset dull, aching or burning pain. If you hit your finger with a hammer, you will experience a sudden pain response followed by a dull pain response. Other types of fibers that transmit touch and vibration exist do not cause pain in most instances. However, these fibers can become hypersensitive and may contribute to your total pain experience. A neuron is an electrically excitable cell in the nervous system that processes and transmits information. Neurons are the significant core components of your brain and spinal cord as well as your peripheral nerves. Neurons are typically composed of a cell body, a dendrite and an axon.. Neurons receive input from dendrites and transmit output via the axon.

Neurons are the building blocks of nerves. In other words, multitudes of neurons are necessary to form a nerve. Nerves that exist outside of your central nervous system are called a ganglion. Your stellate ganglion in your neck is an example. An injection into this ganglion may relieve pain associated with Reflex Sympathetic Dystrophy (now called Complex

Regional Pain Syndrome). Various ganglia may form a plexus. An example of a plexus is your celiac plexus. Sometimes this plexus is blocked with numbing medicine or phenol or alcohol to relieve severe abdominal pain.

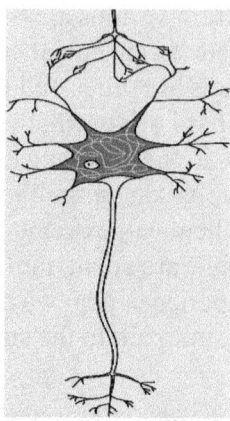

Figure 1. A nerve is composed of neurons. A neuron has one axon that takes nerve signals away from the neuron. The end of the axon communicates with multiple dendrites.

Action potentials generated by the neuron initiate pain signals. If your skin is pinched a mechanical pain receptor generates an action potential. An action potential begins after a depolarization (a change in the electrical activity within the neuron) such that it could cause a membrane transitory modification, making it more permeable to sodium ions more than to potassium ions. Sodium permeability can cause an action potential. Neuropathy generates a local accumulation of sodium channels. This seems to be the basis of neuro hyperexecitably. Calcium channels have also an important role in the cell membrane. Intracellular calcium increase contributes to depolarization processes, through kinase and determines the phosphorylation of membrane proteins that can make powerful the efficacy of the channels themselves.

Following an acute injury, AMPA receptors are stimulated which cause sharp pain. Receptors (areas in the body where biochemicals or drugs attach) are present in the spinal cord are called NMDA (N-methyl-D-aspartate) receptors and cause chronic pain.. When these NMDA receptors are stimulated, pain becomes more severe and this severe painis maintained which implies that the pain does not decrease. The brain is responsible for the suffering associated with pain.

 Pain results in bodily responses especially with respect to the cardiovascular system (heart rate increases, blood pressure increases, renal arteries constrict etc). When pain is severe, the brain can cause the body to increase both the heart rate and blood pressure. Severe pain can also result in profuse sweating as well as nausea and vomiting. There are different types of nerve endings throughout the body. The pain nerve endings become hyper excitable when stimulated by injury, inflammation or a tumor. Occasionally the nerve endings remain irritable even after the painful stimulus has been removed. Pain signals from areas in the body reach the brain by four processes (transduction, transmission, modulation and perception).

Figure 2demonstrates by the arrows that pain signals enter the back of your spinal cord. They cross over to the other side. The pain impulses will then proceed upwards to go to your brain. It is important to know that pain signals can be dampened by structures and chemicals that exist in your spinal cord. Pain signals as mentioned previously are transmitted from the site of injury as action potentials. Electrical and/or chemical activity between the neuron dendrites and axons propagate the axon potentials.

Axons carry pain fibers away from your neuron and direct them to the dendrites of the next neuron until they terminate in your brain or spinal cord. Remember that the axons and dendrites do not touch. They form synapses or clefts between the axon and dendrite. The synapse has chemicals in the axon nerve ending. These chemicals allow communication between the neurons. Drugs are chemicals that can interrupt the communication between the neurons.

You need to understand that pain signals cross to the opposite side from the injury and therefore travel to the opposite side of your brain. Figure 2 demonstrates this concept.

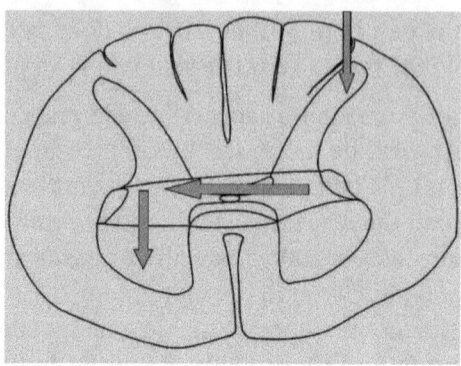

Figure 2. Cross section of spinal cord that demonstrates the site (at the top of the diagram) where pain signals enter your spinal cord, then cross to the other side of your spinal cord and proceed upward to your brain.

It is important to understand four processes in order to understand how your pain can be treated effectively. Transduction is a process where electrical signals originate in the nerve endings throughout your body. These impulses are chemically, mechanically and/or thermally mediated and transmitted to your spinal cord where they can be modulated and then sent to your brain. Tissue injury or disease (including arthritis) cause the body to release biochemicals called prostaglandins.

Prostaglandins themselves do not cause pain. Prostaglandins do however sensitize pain receptors to other chemicals in the body, which facilitate the transmission of pain impulses. Nonsteroidal drugs like ibuprofen decrease the number of prostaglandins produced in your body and may in a decrease in your pain perception. Topical creams such as Ben Gay can decrease the process of transduction at the nerve endings.

Transmission is a process where pain signals are transported to the spinal cord. Nerves in body tissues transmit impulses to the spinal cord. Nerve blocks with anesthetics like Novicaine can interrupt the transmission of pain impulses to the spinal cord. Once pain impulses reach the spinal cord they are modulated or changed by chemicals and nerves that inhibit or lessen the number of pain impulses from going up the spinal cord to your brain. Fibers called internuncial fibers are present within the spinal cord that can decrease pain transmission. The brain can send impulses back to these pain control fibers within the spinal cord to decrease the number of impulses that reach the pain perception center of the brain. This is the basis of hypnosis.

Severe pain however, overwhelms the inhibitory nerve fibers and they essentially become ineffective. Most pain impulses cross over into the opposite side of the spinal cord from where they entered the spinal cord. The spinal cord acts like a transformer to intensify or decrease the intensity of pain impulses. Narcotics and anticonvulsants can modulate pain impulses within the spinal cord. Finally, pain impulses reach the brain where you perceive pain. Be aware that pain signals enter the posterior part of your spinal cord (figure 1) and then cross to the other side and travel upwards to the pain processing of your brain.

Narcotics can "numb" your brain to decrease the effects of the pain impulses on your brain by decreasing the intensity of these impulses. Higher brain centers determine how we respond to a painful stimulus. This explains why an individual can respond differently to a painful stimulus from other individuals (eg. "cry baby, whiner, vs macho man etc.). A chapter describing the anatomy and physiology of pain is not complete without an explanation of the Gate Theory of pain. Melzak and Wall described this theory in 1965. Different types of nerve fibers (both pain and non pain fibers) enter the spinal cord at the same time. Non-pain fibers essentially dilute out the number of pain impulses that enter the spinal cord.

An example of the Gate Control Theory is given by the following analogy. If you can imagine severe pain impulses represented by of multiple black balls going down a sink (analogy to spinal cord). If one adds multiple white balls (neutral non-pain transmitting entities), the number of black balls (pain impulses) is diluted. Therefore less severe impulses reach the spinal cord and the brain. The white balls are non-pain balls and can close the gate (drain) to the number of black balls that go down the sink. To open the gate to more pain impulses, one only needs to decrease the number of white balls going to the hole in the sink. At this time, there are more black balls (pain impulses) available. The gate is now open. As you can see, pain perception in the human body is complex. Because there are many different chemical transmitters and anatomic structures that contribute to chronic pain syndromes, each patient's treatment must be individualized. This is where the art of pain medicine is separated from pure science.

In order to understand pain transmission concepts, you must first become familiar with several biochemicals that are stored in your body that affect your pain signals. In order for you to hurt, pain-producing chemicals in your body tissue must stimulate pain fibers (Alpha-delta and C fibers). In general, the greater the tissue trauma, the more pain transmitting chemicals are produced and the worse the pain.

In medical terminology, a stimulus (pin prick) produces a response (pain perception). When a stimulus such as heat produces, tissue injury chemicals are released at the site of nerve injury, which cause pain fibers to become hyperactive. These chemicals include bradykinin, histamine, substance p, acetylcholine, serotonin and histamine. These chemicals act at the nerve endings and ultimately travel to the spinal cord and brain.

The nerves that conduct pain go to your spinal cord allows pain signals to ultimately reachyour brain. Areas of your body that have many pain receptors include the skin, the outer aspect of bone called the perios-teum, ligaments, joints, teeth and gums and the cornea of your eye. Muscle also contains pain fibers but not as many per square meter (a measure of area) as the previously mentioned structures.

Where the nerves from your body enter your spinal cord, aspartic and glutamic acids are produced. These acids increase pain impulse generation. NMDA may also be produced. GABA (gamma-aminobutyric acid) in the spinal cord on the other hand, decreases the number of pain impulses that reach the brain. GABA inhibits pain impulse transmission   Norepinepherine and serotonin are two more chemicals in the spinal cord which attenuate the number of pain im-pulses which reach your brain. The brain and spinal cord regulate pain by the production of naturally occurring narcotic-like substances that decrease pain transmission in specific areas of the brain. These narcot-ic-like drugs are called enkephalins, dynorphins and beta-endorphins. Some of these substances also decrease pain transmission in the spinal cord. Enkephalins are located in areas of the brain related to pain modulation.

Enkephalins inhibit pain at the spinal cord level. Enkephalins bind to narcotic receptors. When the narcotic receptors are activated, they inhibit pain signals. Dynorphins exist in both the brain and spinal cord but are more prevalent in the brain. Like enkephalins these substances bind to narcotic receptors in the brain and spinal cord. Pain impulses that enter your spinal cord cross over to the other side and then progress upward to your brain. The natural beta-endorphins in your body exhibit morphine-like activity. They work like morphine to decrease your pain. Following injury or stress these endorphins are released into the blood stream. The effects of beta-endorphins are similar to morphine. Beta-endorphins like narcotics can cause respiratory depression, constipation, euphoria, tolerance and physical dependence. The exact biochemical actions of all of the substances mentioned are complex.

For a more detailed explanation of the actions of these substances one should consult a pain medicine textbook. The purpose of this chapter is to emphasize the multiple substances that can generate the transmission of pain signals. This is furthermore the reason why there are so many medications available for the management of your pain. This is also the reason why your physician may prescribe multiple medications for the management of your chronic pain.

With respect to tissue and nerve ending biochemicals, neurotransmitters and pain transduction, your physician may recommend a skin (topical) cream to decrease the transmission of pain signals to your brain. Red pepper cream decreases the pain generator called substance P. An example is Zostrix cream. Menthol containing creams (Ben Gay) also decrease pain over muscles and joints. Non-steroidal anti-inflammatory drugs (NSAIDS) decrease the production of prostaglandin that can sensitize your body to pain mediators. Examples include Advil and Celebrex.

Remember that prostaglandins sensitize pain nerve endings to pain producing tissue chemicals. Antidepressant drugs like Elavil or Prozac decrease pain by increasing norepinephrine and serotonin in the spinal cord. As previously mentioned, these two substances decrease the number of pain impulses that reach the pain perception areas of the brain. Anticonvulsant drugs like Gabitril (tiagabine) affect GABA and by enhancing GABA blood levels which in turn decreases the number of pain signals in your spinal cord that can go to your brain. Narcotic drugs also decrease pain impulse conduction in both the spinal cord and brain. Injections of numbing medicine (local anesthetics with steroids) can decrease pain in muscle and nerves in the arms, legs and the trunk of the body.

Epidural steroid injections can decrease pain in nerves that are buried deep within the spine. As you see there are multiple biochemical sources of pain and in many instances your physician may elect to prescribe multiple medications with good reason. Remember that each of these medications can have side effects that will be discussed in a later chapter. There is an area of your brain that represents an area where you process pain signals. This area detects tissue injury and is a protective mechanism to alert you that something is wrong. A burn of the palm of your hand alerts your brain that tissue injury is occurring and initiates a reflex in your spinal cord to have you immediately remove your hand from the hot object. Without pain interpretation in your brain, you could sustain multiple bodily trauma and have no knowledge of its occurrence.

The different dimensions of pain perception have been shown to depend on different areas of your brain. In contrast, much less is known about the neural basis of pathological chronic pain. Patients may report combinations of spontaneous pain and allodynia/hyperalgesia-abnormal pain evoked by stimuli that normally induce no/little sensation of pain. Modern neuroimaging methods (positron emission tomography (PET) and functional MRI (fMRI)) have been used to determine whether different neuropathic pain symptoms involve similar brain structures.

PET studies have suggested that spontaneous neuropathic pain is associated principally with changes in thalamic activity and the medial pain system, which is preferentially involved in the emotional dimension of pain. Not only are there areas of your brain where you perceive pain but there are areas that are responsible for suffering as well. An area of your brain called the amygloid is associated with fear. Animals who have had their amygloid areas excised do not exhibit fear. Fear, suffering and pain are in different areas of your brain but these areas are connected to each other. These interconnections ultimately can communicate with areas of your brain that control your heart rate and respiratory rate as well. If you have severe pain you may sweat profusely in addition to having increases in your heart and respiratory rates. As you can see, severe pain can have adverse physiologic effects on your body.

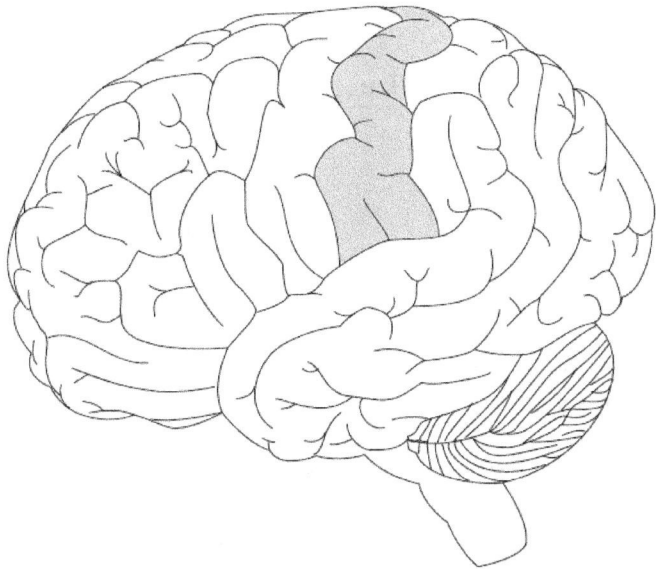

Figure 3. The brain areas (shaded) where pain signals are processed.

3. Methods of Pain Relief

This chapter presents a general overview of modalities available to treat your pain.   These modalities will be discussed individually in later chapters in this book.   The ability to relieve pain is very variable and unpredictable, depending on the source or location of pain and whether it is acute or chronic. Pain mechanisms are complex and have peripheral and central nervous system aspects. Therapies should be tailored to the specifics of the pain process or processes in the individual patient.

Anesthesiologists do nerve blocks and prescribe narcotic medications. Physical medicine and rehabilitation specialists prescribe physical therapy exercises and modalities using heat and cold to relieve pain. Neurologists control pain with anti seizure medications.   Psychiatrists decrease pain with antidepressant drugs. Psychologists manage pain with psychological counseling, hypnosis and biofeedback.   Rheumatologists administer steroids into painful joints.

Physical therapists will rehabilitate your body.   Chiropractors manipulate your spine to control your back and neck pain by realigning your spine and taking pressure off your nerves. Surgeons operate on nerves, joints and discs to lessen the suffering of pain. In many chronic pain states a multidisciplinary approach using all or most of the mentioned pain specialists is used to manage complex pain problems such as reflex sympathetic dystrophy which will be described in a later chapter. Which modalities actually work? Each of these modalities can help you with your pain. In many instances a multidisciplinary pain center can provide you with the most benefit.   You should investigate what therapies are scientifically sound. You should have some basic knowledge about scientific studies that have been.

Suppose your neighbor tells you that you need procedure "X" to manage your pain. Her doctor does this procedure and your neighbor says that her doctor will do it on you. You should take the time to examine studies that recommend this procedure as being wonderful and revolutionary. For example injection "X" is reported in a newspaper advertisement to relieve low back pain in 80% of patients. WOW! This must be a great procedure. You do some investigating. The results are placed in the following table.

	Pain Relief (%)	No Pain Relief (%)
Group I (100 patients)	80 %	20 %

Table 1. Pain relief with new pain device.

What is missing? You need to know how many individuals with the same pain symptoms who did not have this procedure had relief of their pain. On the other hand, you could look at a group of patients with the same treatment who received another drug or substance. This is called a control group. Now compare injection "X" treatment with what is referred to as a sham treatment or a control treatment. . A sham treatment is essentially a placebo treatment. Now examine the results comparing injection "X" with the placebo. The placebo is essentially better than injection "X". A good study will always have a comparison group. Do not be fooled by testimonials or by studies that have no comparison group. Remember that statistics can be manipulated if there is no control group.

	Pain Relief (%)	No Pain Relief (%)
Group I (100 patients)	80 %	20 %
Group II (100 patients)	81 %	19 %

Table 2. The natural course of the pain is similar to injection X.

You can see now that there is no difference in the groups that received the treatment compared to individuals who did not receive the treatment. You should not be fooled by claims of success that have no merit.

Nerve blocks are frequently used to manage pain. Local anesthetics like Novocain used in combination with steroids are deposited near the nerves or tissues that are responsible for chronic pain. These anesthetics stop pain production by numbing the nerves responsible for your pain. The steroids decrease the irritability of the pain producing nerves. In many instances these blocks will break the pain cycle. Many patients respond to drug therapy. Mild and strong narcotics are prescribed depending on the severity of your pain. Pain patches with either a local anesthetic or a strong narcotic exist which give a patient sustained pain relief. Long acting narcotics (Oxycontin) exist which decrease the need for frequent drug dosing.

Antidepressants and anticonvulsants modulate pain transmission in the spinal cord. Muscle relaxants decrease muscle spasm that can significantly decrease your pain. Non-steroidal anti-inflammatory drugs alleviate pain by decreasing tissue inflammation. The Chinese have used acupuncture for over 2000 years. This method of pain relief consists of placing small needles into the skin and muscles over the body. The needles stimulate larger nerves that go to the spinal cord and release endorphins and enkephalins. These substances decrease the number of pain impulses that go to the brain. Chiropractic therapy consists of manipulating the spine by a physician trained in safe spinal manipulation.

A chiropractor aligns the spine. This maneuver takes pressure off the nerves coming off the spinal cord that decreases pain conduction. Psychologists may help control pain with hypnosis or biofeedback. Hypnosis helps activate the nerves in the spinal cord that block pain signals from traveling up your spinal cord to your brain. Biofeedback uses a machine to enable a pain patient to relax painful muscle. Physical therapists administer modalities that provide heat and cold to your muscles and ligaments. Some therapists do massage therapy that relaxes painful muscles. Electrical stimulation applied to the body decreases pain in a variety of painful conditions. The device is called a TENS (Transcutaneous Electrical Nerve Stimulator) unit. This instrument is battery powered.

A TENS unit stimulates endorphin and enkephalin production in your spinal cord. Neurosurgeons can place implantable devices in your body to control severe pain. One device is called a narcotic pump. This device gives a drop of morphine or other strong narcotic into the fluid around the spinal cord. The other surgically implanted device is called a spinal dorsal column stimulator. A wire is attached to a battery source that is placed in your body. The wire is placed parallel to your spinal cord. This stimulation releases endorphins and enkephalins within the spinal cord. As one can see many modalities are available to patients for the control of their chronic pain. Which one is right for you? The proper treatment for your pain will depend on the severity of your pain as well as your physical and mental health status.

When considering a modality to relieve your pain, you should be aware of another important concept. Evidence-based medicine (EBM) is an attempt to more uniformly apply the standards of evidence gained from scientific medical studies to certain aspects of medical practice. Specifically, EBM seeks to assess the quality of medical evidence relevant to the risks and benefits of treatments (including lack of treatment). According to the Centre for Evidence-Based Medicine, "Evidence-based medicine is the conscientious, explicit and judicious use of current best evidence in making decisions about the care of individual patients.

EBM, however, seeks to clarify those parts of medical practice that are in principle subject to scientific methods and to apply these methods to ensure the best prediction of outcomes in medical treatment, even as debate about which outcomes are desirable continues". EBM requires clinical expertise, but also expertise in retrieving, interpreting, and applying the results of scientific studies and in communicating the risks and benefits of different courses of action to patients.

Evidence-based medicine/healthcare is looked upon as a new paradigm, replacing the traditional medical paradigm that is based on standard of care authority (which is common practice in your state). It is dependent on the use of randomized controlled trials, as well as systematic reviews (of a series of trials) and meta-analysis, although it is not restricted to these. There is also an emphasis on the dissemination of information, as well as its collection, so that the evidence can reach clinical practice. It therefore has commonality with the idea of research-based practice. Having considered the extent to which EBM represents a departure from basing medical decisions on customary practices, there may be a change in the extent to which medical custom remains prevalent in the legal standard of care analysis. A problem ensues in the custom-based standard of care that courts have traditionally used to determine medical malpractice liability. If you are injured from a medical procedure, should the court use the old procedure as the standard of care or the new procedure that is supported by evidence based medicine? For these reasons, current standard of care analysis is potentially inconsistent with the practice of EBM.

This concept is important because traditionally, local customs established the standard of care in medical malpractice actions. Under a custom-based standard, practicing in accordance with accepted practice generally decreases medical liability. Physicians have in the past needed only to conform to the "customs of their peers."

As you can see, EBM can cause some legal problems if you are injured from a procedure in a certain medical community. If the standard of care was to do unsafe medicine that EBM demonstrated to be unsafe you need a legal consultation. For example, it used to be acceptable to put steroids in your spinal fluid for the management of back pain. It is no longer the standard of care in most communities. What is your recourse if you had this procedure and developed complications? Complications could take months to years to develop in some instances. You need to realize that EBM practices may differ from customary care in your community. For this reason, you should not attempt to handle your own case. Instead, you should seek the advice of an attorney.

4. Cost of Pain Management

Approximately forty-eight million Americans suffer from chronic pain. According to a previous Wall Street article, Americans spend over 100 billion dollars on pain care. Over one third of all adult Americans suffer from chronic pain. Over the counter annual analgesic costs amount to three billion dollars. Chronic pain is a prevalent and a costly problem. One must assess the clinical effectiveness and cost-effectiveness of the most common treatments for patients with chronic pain.

Untreated or under treated chronic pain costs health plans and their purchasers a lot of money. It afflicts 40 million or more Americans, with a cost of nearly $100 billion a year in direct medical costs and indirect costs, such as lost productivity and workers' compensation, according to the American Pain Society. Most primary care physicians are not adequately treating pain, primarily because they are not trained to do so and because they are afraid of litigation and regulatory restrictions. Chronic pain with conservative care such as medications, physical therapy, chiropractic etc. costs North American adults an estimated $10,000 to $15,000 per person per year. Furthermore, estimates of the cost of pain do not include the nearly 30,000 people that die in North America each year due to non-steroidal anti-inflammatory drug-induced gastric lesions.

Because patients usually have to pay a significant portion of their bill for fees charged for procedures done on them as well as for medications prescribed, representative published studies that evaluate the clinical effectiveness of pharmacological treatments, conservative (standard) care, surgery, nerve blocks and pain rehabilitation programs must be examined and compared. When you're in pain, the last thing you want to think about is the cost of getting the relief you deserve.

Your physician can help. The cost-effectiveness of various treatment approaches must also be considered. Outcome criteria of a particular treatment should include the prevalence of pain reduction. Patients prior to consenting to any treatment modality should evaluate medication use and health care consumption. In addition to clinical effectiveness, the cost-effectiveness of conservative care, surgery and nerve blocks must be compared. There are limitations to the success of all available treatments. Chronic pain programs are in many instances in need of rehabilitation. Although hard statistics regarding such programs are difficult to obtain, one frequently hears of programs closing down or modifying their treatment protocols to meet their own survival needs rather than meeting the needs of the patients they serve.

After rapid growth during the 1980s and through the mid-1990s, the number of inpatient chronic pain management programs actually declined. Concurrent with the decline in intensive programs is the rise of procedural interventions and medications, which receive a great deal of support from hospitals and pharmaceutical companies. The use of muscle relaxants for patients appears to be increasingly prevalent when compared with teaching relaxation techniques, and implanting a device is more lucrative than giving patients guidance or advice.

Healthcare specialists have to determine whether this apparent shift in treatment emphasis away from rehabilitation is a healthy development for the patients they serve.   Many hospitals encourage physicians with minimal or no training to open pain clinics in their facilities. They can charge facility fees of $1000.00 or more for a procedure like an epidural steroid injection. Your physician doing the procedure may own shares in the center or hospital. As a result, every time that a physician schedules a procedure in the hospital he or she makes a share of the profit. The sad fact is that this behavior is legal.

Medical procedures, such as trigger-point injections, sympathetic nerve blocks, and epidural steroid injections, are rated as significantly less helpful , less invasive modalities; despite their considerably higher average cost. Research is needed to identify which patients are most likely to benefit from the available treatments and to study combinations of the available treatments since none of them appears capable of eliminating pain or significantly improving functional outcomes for all treated. The cost of chronic benign (non-cancer pain) spinal pain is large and is increasing. The costs of interventional treatment for spinal pain were at a minimum of $13 billion (U.S. dollars) in 1990, and the costs are growing at least 7% per year. The interventional medical treatment of chronic pain costs $9000 to $19,000 per person per year. It should be understood that only a small percentage of patients receive long-term relief with these procedures. You should inquire about the cost of any treatment before agreeing to have the procedure done. You also need to contact your insurance carrier and ascertain if your treatment is covered under your insurance plan.

Before consenting to a potential expensive therapy do research. Check the Internet for a description of the procedure and if it is effective. Ask your physician what the cost is for the treatment in question. Is there a surgery center charge in addition to the physician fee? You should ask your physician if there is an extra fee for sedation if you have to have a nerve block. Some centers bundle this fee with the facility charge.

Your treatment for your back pain may only cost $3.00 for a bottle of aspirin. However, you need to calculate and add your chiropractor fees or the costs of a massage. What about the housekeeper you had to hire to do the housework that you can't do while you were disabled? What about the heating pad? What does it cost you if you need to cut down on the hours you work, or quit your job because of your pain? Pain costs can affect your family if your spouse has to take off work to help you.

Pain treatment can be expensive for you but a question arises concerning the cost of you not receiving pain management. Pain patients frequently miss work because their pain is too severe to allow them to work. As a result business productivity can be decreased. Lost workdays also cost businesses money. The overall cost of pain amounts to over sixty billion

dollars per year. When deciding on whether or not to get pain treatment the following are recommended: call your medical insurance company and find out what treatments your insurer will cover.

You need to ascertain if you have to pay co-payments and how much those costs are. You also need to know whether the balance of the charge beyond the co-payment will be billed to your insurer, or whether you have to pay in full when the service is provided and then request reimbursement from your insurer. You should also ask your insurance company if it will pay for alternative medicine therapies like accupuncture.

If you could look at your past 12 months and calculate how much time that you lost from work because of your pain and compare these figures to your cost of receiving care, you can estimate the actual cost of your treatment. For example, if an epidural steroid injection costs $500.00 but saves you 3 days of lost work time at $200.00/day, you actually save $100.00. If an injection keeps your activities of daily living normal, then one would expect the treatment to be cost effective if the treatment itself is effective.

Professional fees charged by physicians may differ from doctor to doctor. In some areas of the United States, physicians will post their professional service fees somewhere in their office so that you are informed of the costs before you consent to a procedure. Before you seek medical care, you should call several pain management offices to ascertain professional fees. Your primary care physician will then refer you to a physician that you chose.

5. Pain Assessment

Pain is a subjective experience for which there are no objective biological markers. There is no objective measurement of pain. Self-report is considered the most accurate and appropriate pain assessment method as family members and caregivers often underestimate a patient's pain. As researchers work toward developing better treatments for chronic pain they will need adequate ways to assess pain and its effects on life. Patients should be asked to rate their pain both to better understand its severity as well as to give a baseline assessment to determine changes in the level of pain after treatment. Pain is the most common reason why a patient visits a physician.

Pain is a complex entity and is affected by tissue injury but also by previous pain experiences, as well as by your psyche (anxiety and depression) and by any pending lawsuits. The transition from acute pain following an injury to chronic pain cannot be explained.   The entity that we refer to as pain cannot be touched or felt by the treating physician. As a result, you must give your doctor an accurate assessment of your pain. Many chronic pain patients have seen a variety of physicians before seeing a pain medicine specialist. Many patients become frustrated and sometimes feel that no one believes that they truly hurt.

A patient's pain may not correlate with findings or lack of findings on a physical examination and X- rays. A patient's pain can only be adequately diagnosed after the treating physician has done a good history and physical examination. A clinician cannot feel your pain. As a result, you will be given papers to fill out with questions to answer. Questions asked will include a history of your pain, a pain diagram and occasionally a depression and/or activity assessment.   Significant depression can worsen your pain.

Several different techniques are available for your doctor to use in determining your level of pain. Commonly used techniques include verbal, visual, and psychological 'tests. Both you and your doctor are responsible for documenting and recording trends in the intensity and frequency of your pain. This information tells each of you whether your pain has really improved or whether it has worsened.

Charting your pain levels in a diary will help your doctor see your long-range (weeks-months) pain trends, which are ultimately more important than your day-to-day pain trends. One important factor used to assess your progress is your activity. Did you return to work after taking off for a painful episode? Are you playing tennis etc.? A pain assessment provides a baseline for your doctor to assess any therapy or medications you are currently taking, and it also helps your doctor prescribe future therapy methods. Your doctor also needs to be able to determine how much disability you have in order to prescribe the appropriate types of therapy for you.

Your doctor will depend on you for accurate and reliable answers to questions about the pain you experience. Because pain involves many aspects such as sensory, emotional, and behavioral factors, it is difficult to measure the amount of pain you feel based on one single parameter. The choice of a pain-assessment test depends on the needs of both you and your doctor. A functional evaluation, such as reports of your daily activities, must be included in your assessment. If your doctor does not ask about your daily activities, voluntarily tell him your further limitations with respect to work, recreation, dressing, fixing meals, and any other daily activities. Progress continues to be made in developing pain-assessment tools. You or your doctor should not oversimplify your pain assessment. The objective reports you are able to give, as well the observations your doctor is able to make about your behavior, are important to accurate pain management decisions. Because pain is subjective and can be observed only by you, it is important that the reports of your pain levels come from you. This will give your doctor a more accurate measurement of the type of pain you are experiencing.

For example, if you just complain of a toothache, your doctor will have almost no way of knowing how severe your pain condition is. There is no general consensus among pain medicine doctors as to the best test for the measurement of pain. An ideal test for the assessment of pain must bring together experimental as well as clinical knowledge.

One way of assessing your pain is to use a numeric scale. This is the simplest method for attempting to measure your pain. During this test, you are asked to rate your pain on a scale of 0 to 5 or to use words such as "none," "slight," "moderate," or "severe." This assessment is also a quick, simple, and reliable way to evaluate the effectiveness of any medications you are taking to manage your pain.

On the numeric scale, 0 equals no pain, 1 equals mild pain, 2 equals moderate pain, 3 equals distressing pain, 4 equals horrible pain, and 5 equals excruciating pain confining you to bed rest. This method is easily understood and may be helpful in guiding the treatment plans your doctor creates for you. Another type of verbal scale asks you to rate your pain on a scale of 1 to 10, with 1 being equivalent to pain that is barely noticeable and 10 relating to excruciating pain. A verbal numeric scale is easily understood. All you have to do is choose a number to represent your level of pain.

Another method used by some doctors is a pain diary. This is a descriptive report you keep to assess your pain. The pain diary shows a written account of your day-to-day experiences. It can be used to help diagnose the causes of your pain. The value of the pain diary is that you and your doctor can monitor your day-to-day variation of painful states and your response to therapy. You need to keep a diary of your pain patterns when you are sitting, standing, and lying down. You should also record your sleep patterns and sexual activity. You also must note the amount of pain medication you are taking and whether it lessens your pain. Because pain can interfere with eating patterns, keep a diary of the amount of food you eat and at what time you ate. Be sure to include any types of recreational activities and whether your pain felt better or worse afterward.

Pain drawings offer a visual way to evaluate your pain. You will be asked to shade in areas on a human figure outline that correspond to the areas of your pain. The drawing will help your doctor determine where your pain is coming from and how widespread it is on your body. Over time, your pain drawings can be compared to show the changes of your pain and how you are responding to therapy.

After observing your behavior, your doctor may classify your pain behavior by using the following four-class system: Class 1 consists of patients with low physical injury but high levels of abnormal behavior patterns related to their pain. Class 2 consists of patients with lower physical injury and low behavior pattern abnormalities. Class 3 consists of patients with significant tissue injury in addition to high behavioral pattern abnormalities. Class 4 consists of patients with a high tissue injury and normal behavioral patterns.

A McGill pain questionnaire is a method for assessing pain psychologically. A McGill pain questionnaire gives a multidimensional pain score. You are given 20 word sets that describe a different dimension of your pain. You are asked to select words relevant to your pain from each of these 20 sets. For example, one set includes the words "jumping," "flashing," and "shooting." Another set includes the words "tingling," "itching," "smarting," and "stinging." You circle the word that relates closest to the pain you feel throughout the 20 word sets. The McGill pain questionnaire consists of four different parts. The first part consists of a human figure drawing on which you are instructed to mark the location of your pain. The second part is the pain-rating index that contains 78 words divided into 20 groups. Each set contains up to six words. Five of these groups describe tension or fear. Each word is assigned a value according to its position within a subclass. The third part of this test asks additional questions about prior pain experiences, as well as the location of your pain and current usage of pain medications. The fourth part consists of a present pain intensity index. This aspect of the test requests a pain score from 0 to 5 with word descriptors such as no pain, mild pain, discomforting pain, distressing pain, or horrible and excruciating pain. These words also are assigned different values. All the values are added to obtain a total score. All the scores are then evaluated to attempt to assess your total pain experience.

There is also a short form of this test that has been developed. This questionnaire contains fewer words and categories than the long form. This test is sensitive to evaluations of reduction in pain experiences. This test is more useful for rapid evaluation of data following procedures or surgery.

Your physician will ask many questions or give you a form to fill out which addresses the location of the pain, the intensity of the pain, the type of pain (e.g. burning, dull or sharp). Your physician needs to know what makes your pain worse and what relieves your pain. The severity of a patient's pain is the hardest entity for a physician to know because your physician is not experiencing your pain. One way of assessing the severity of your pain is to assess your blood pressure and pulse. These are called hemodynamic parameters. They can be elevated when you are experiencing severe acute pain. These parameters are not helpful however, in chronic pain. A patient may complain of a severe toothache but have no increase in the blood pressure and heart rate. On the other hand, a patient with invasive cancer with excruciating pain will probably have significant increases in heart rate and blood pressure.

 Other pain self-assessment scales are available. The horizontal visual analog scale consists of a 10 cm line anchored by two extremes of pain: no pain and extreme pain. Patients are asked to position a sliding vertical marker to indicate the level of pain they are currently experiencing; pain severity is measured as the distance in centimeters between the zero position and the marked spot. The vertical visual analog scale is similar to the prior scale but is presented vertically, and the line is replaced by a red triangle with its summit facing downwards (no pain=0) and its base at the top (maximum pain=10). The faces pain scale consists of a line drawing of seven faces that express increasing pain (no pain=0, maximum pain=6).

 Many pain physicians use Verbal Analogue Scales (VAS) to assess the severity of your pain. A scale of 0 to 10 can be used. A score of 0 indicates no pain while a score of 10 represents the worst imaginable pain. This scale is simple and is easy to be understood by the patient. This is similar to the previous scale with the exception that a patient makes a mark on a ruler with marks from 0 to 10. Your pain ratings will be kept in your patient chart and the scores will be compared with each visit. These scales serve as a gauge of your progress to treatment.

Some doctors want to have their patients keep a pain score diary on a daily basis. Your doctor may want to know the relationship between your pain and your activity such as sitting, standing and walking. Since sleep deprivation can worsen your pain, your doctor may ask you about your sleep pattern on each visit. Your doctor needs to know if you can do normal activities as well as whether or not you can work. In order for a physician to formulate your treatment plan, your physician needs to know if the pain is localized or referred. Localized pain is confined to one area such as a knee joint.

Referred pain begins in one area and travels to other areas of your body. An example is pain referred from the heart during a heart attack to the left arm. A pain medicine specialist also needs to know if your pain is superficial or deep. An example of superficial pain is that from a fingernail or from the skin as seen following sunburn. An example of deep pain is that from a structure deep within the body such as a diseased appendix or gallbladder. One treatment may only require medications while the other may require surgery. As you can see the treatment for each of these types of pain will differ. Superficial pain may be treated by analgesic creams or by pills.

Deep pain as seen in cancer patients may require needle injections or surgery to control the pain. Referred pain such as pain originating from the heart may require a surgical procedure or may require only medications. Localized pain can be treated with an injection of anesthetics that can numb the painful area. An example is an injection of a knee joint. You should begin to understand that the treatment of pain could be complex and on occasion be frustrating for both the patient and physician.

Before a physician begins pain treatments, a complete examination must be done. Strength, sensation and reflexes must be evaluated. Skin temperature should be noted as well as range of motion of the neck, back and extremities. A patient's mental status must be evaluated. Vital signs and the way a patient walks should be recorded. Swelling, skin color and hair loss should be appreciated. Loss of sensation to touch can be evaluated by rubbing a fine brush over the skin. Strength and coordination are evaluated as part of the neurological examination. Following your examination your physician will probably want to order diagnostic studies. Plain X rays are used to diagnose bone abnormalities of the spine. A CAT scan is used to define bone abnormalities, as well in more detail than a plain X ray.

You may ask why your doctor cannot diagnose the source of your pain by looking at your laboratory or imaging studies. These studies only relate that there is an abnormality in your pathology. You need to know that if you have abnormal tests, it does not mean that you hurt. If you have degenerative disc disease noted on X ray, you do not have to have pain. The X-ray is only a picture. For example, you may have a picture of a telephone. Is it ringing? There is no way to tell from looking at the picture that you are in pain. The same is true with respect to imaging studies and laboratory tests.

An MRI is useful for the diagnosis of soft tissue pathology such as nerve compression or muscle shrinkage, disc herniations and tumors. However, the MRI does not tell if you hurt. You can have a disc herniation but not experience any pain. A myelogram is the injection of dye into the fluid that surrounds the spinal cord. This test can visualize a disc herniation or compression of a nerve coming off of the spinal cord. A bone scan is the injection of dye into your vein followed by a series of pictures of your skeleton that are recorded by a scanner. This test can detect arthritis and trauma to bone as well as tumors. An EMG is the evaluation of your muscle using needle electrodes. This test can detect muscle pathology.

Nerve conduction studies evaluate abnormalities of the transmission of electrical impulses through nerves. Blood tests can detect rheumatoid arthritis and some medical diseases such as liver disease. A urinalysis may detect kidney pathology. As you can see the diagnosis and treatment of chronic pain can be difficult and challenging. Because pain can have multiple causes, a physician must treat you as a whole and not just your area of pain. For example, you may have pain in the bottom of your feet. This pain may be the result of a vitamin deficiency. Your doctor could give you some numbing medicine (e.g. a Lidoderm patch) for your feet that may decrease your pain for a short time. However, if your vitamin deficiency is corrected, your pain may completely resolve.

The verbal rating scale is a simple, commonly used pain rating scale. To complete it, subjects select one of six descriptors that represent pain of progressive intensity: none, mild, discomforting, distressing, horrible, or excruciating. Another scale is a modified 21-point Box Scale. The scale has a row of 21 boxes labeled from 0 to 100 in increments of five. The 0 is labeled "no pain," while the 100 score is labeled "pain as bad as it could be."

Pain can be extremely difficult to assess in elderly patients. It can be difficult for elderly patients to give a numeric representation of their pain. They should be asked to verbally describe their pain as none, some or severe. Pain may be particularly difficult to identify in cognitively impaired individuals as it can manifest itself atypically as agitation, increased confusion, and decreased mobility. In many clinical settings, pain is not assessed in demented patients due to reliability concerns. In particular, self-assessment is rarely attempted. Furthermore, when pain is evaluated in severely demented patients, the nursing staff routinely uses observational scales. These assessments include vocalizations, facial expressions, and body language.

Pain is also difficult to diagnose in infants and children. Children less than two years of age will report that they hurt but have difficulty localizing the exact area where they hurt. A child with a finger injury may just report pain but not identify the finger as the source of pain. This is because the brain has not developed enough to be able to distinguish generalized pain from localized pain. As a result infants and young children must be observed. Crying, facial expression and blood pressure and heart rate can be good indicators that pain is present.

6. Diagnostic Tests

This chapter describes the various diagnostic tests, which may be ordered by your physician. These tests should be ordered to confirm a doctor's clinical impression. Laboratory tests check a sample of your blood, urine or body tissues. Your doctor analyzes the test samples to see if your test results fall within a normal range. The tests use a range because what is normal differs from person to person. Some laboratory tests are precise, reliable indicators of specific health problems. Others provide more general information that simply gives doctors clues to possible health problems. Information obtained from laboratory tests may help doctors decide whether other tests or procedures are needed to make a diagnosis. The information may also help your doctor develop or revise a patient's treatment plan.

All laboratory tests are generally used along with other exams or test such as MRIs, X-rays, EMGs etc. The doctor who is familiar with their patient's medical history and current condition is in the best position to order and to explain test results and their implications. Patients are encouraged to discuss questions or concerns about laboratory test results with the doctor. Two common tests that you should be familiar with are the complete blood count and the blood chemistry tests. A complete blood count measures the levels of different types of blood cells. By determining if there are too many or not enough of each blood cell type, a CBC can help to detect a wide variety of illnesses or signs of infection. A blood chemistry test measures the levels of certain electrolytes, such as sodium and potassium, in your blood. A C reactive protein and erythrocyte sedimentation rate teat may be useful in the diagnosis of rheumatoid arthritis or other inflammatory disease.

Doctors order urine tests to make sure that your kidneys are functioning properly or when they suspect an infection in your kidneys or bladder. This is important if you are taking a medication like a nonsteroidal anti-inflammatory medication that can affect your kidney. A urine test can be done in the doctor's office or even at home. It's easy for toilet-trained kids to give a urine sample since they can urinate in a cup. In other cases a catheter (a narrow, soft tube) can be inserted through the urinary tract opening into the bladder to get the urine sample.

Tylenol (acetaminophen) can cause liver damage if you take too much (more than 4000 mg per day). Liver function tests ascertain how your liver is working and helps diagnose any sort liver damage or inflammation. Your doctor may order one when looking for signs of a viral infection or liver damage from other health problems. On occasion, blood tests may be done to determine that you do not have a bleeding problem such as hemophilia. Aspirin can cause bleeding by decreasing the ability of your blood to clot. Before doing a nerve block it is prudent to know if your blood will clot in a normal time. Otherwise a needle can result in significant bleeding.

Plain X rays can be done in a physician's office (Figure 1). X rays can assess bone-joint arthritis. X rays can diagnose degeneration of your discs. Your bone alignment (do the bones line up with each other?) can be assessed as well. Bone fractures can also be identified. You should be aware that you are subject to radiation exposure with this diagnostic test. If you have the possibility of having osteoporosis, your physician may order a DEXA (dual energy x-ray absorptiometry) that is a specific test for the diagnosis of osteoporosis. A Computed Tomography (CT scan) allows a physician to assess a disc in your back as well as arthritic changes affecting the bones in your neck and back.

A CT scan of your head can be useful for the diagnosis of a bleeding injury to your brain following trauma to your head. Patients receive radiation exposure with this test. Myelography or a myelogram is primarily of use when surgical therapy is planned. A dye is placed in the fluid that surrounds your spinal cord. An image is formed which tells a physician that a nerve coming off your spinal cord is compressed or not compressed by a disc herniation.

An image does identify painful areas of your body. An image demonstrates abnormal anatomy that could be an area of pain generation.

Degenerative disc disease noted on an X- ray for example does not imply that you have a disease or are supposed to have pain. This entity is a normal aspect of aging. Therefore, you should not be alarmed if your doctor tells you that you have degenerative disc disease. The same is true if you are told that you have a disc herniation. Not every disc herniation causes pain and not every disc herniation requires surgery.

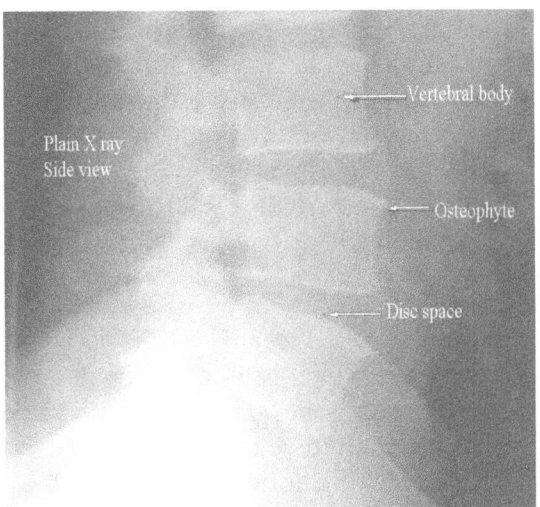

Figure 1. Plain X ray of a low back taken from the patient's side.

Ultrasound is another valuable diagnostic tool. Though ultrasound tests are typically associated with pregnancy, doctors order ultrasounds for different reasons. For example, an ultrasound test can be used to look for collections of fluid in your body, or for problems with your kidneys. An ultrasound is painless and uses high-frequency sound waves to bounce off organs and create a picture. A special jelly is applied to the skin, and a handheld device is moved over the skin. The sound waves that come back produce an image on a screen. Computerized axial tomography is a specialized x ray. CAT scans are a kind of X-ray, and typically are ordered to examine for pathologies such as appendicitis, internal bleeding, or abnormal organ growths. Tomography in which computer analysis of a series of cross-sectional scans made along a single axis of a bodily structure or tissue is used to construct a three-dimensional image of that structure. The technique is used in diagnostic studies of internal bodily structures, as in the detection of tumors or brain aneurysms. A scan is not painful.

A scan may require the use of a contrast material (a dye or other substance) to improve the visibility of certain tissues or blood vessels. The contrast material may be swallowed or given through an IV. CAT scans consist of a highly sensitive x-ray beam that is focused on a specific plane of your body. As this beam passes through your body, it is identified by a detector, which feeds the information that it receives into a computer. The computer then analyzes the information on the basis of tissue density. Generally a CT is preferred where bone details necessary (long bones like your arm or leg, spine, skull), while a MRI produces much better soft tissue details (brain, spinal cord etc) CT scans are useful for examining body cavities (thorax, abdomen, pelvis) for calcium deposits, cysts, and abscesses.

With some diseases, either a CT scan or MRI is commonly ordered. Spinal stenosis, for example which is a bone growth around your spinal cord or around the holes in the bones of the low back and neck where the nerves from the spinal cord exit to your extremities and is usually seen in individuals over 50 years old. Stenosis can compress the nerves resulting in pain and numbness in the extremities. Because of numbness on the bottom of your feet, you may have difficulty with balance. A CT scan or MRI can identify this pathology. Magnetic Resonance Imaging (MRI) is done by utilization of a magnetic field that is applied around your body. MRIs use radio waves and magnetic fields to produce an image.

MRI's (Figure 2) are often used to look at bones, joints, and the brain. Contrast material is sometimes given through an IV in order to get a better picture of certain structures. Nuclei within your body cells with an odd number of protons orient themselves with the magnetic field. The MRI scanner applies a certain amount of energy and the nuclei assume a new orientation with respect to the magnetic field. This energy is removed and the nuclei emit energy as they reorient in the magnetic field. The energy emitted is detected and displayed as an image. The MRI involves no radiation.

Magnetic resonance imaging provides a picture of your soft tissue that may be better than the CT scan. A MRI cannot be done if you have certain metals in your body or a heart pacemaker or a defibrillator. Magnetic resonance imaging allows visualization of your discs, spinal cord and cerebrospinal fluid. A MRI can be used with a contrast dye to identify an extruded disc, infection or tumor.

Plain X rays give physicians images in a front to back plane. A side-to-side plane and a oblique view are helpful in diagnosing your possible causes of your pain. On the other hand, a CT and MRI (figure 2) image shows slices of the body as well as a three hundred and sixty degree image of a defined section of the body.

Images only show pathology. They do not show pain. Pain is a subjective experience. If you view a photograph of an old scratched and dented telephone, you have no idea if it is ringing or not. You have no idea if it works. The same is true with an X- ray image. An abnormal X ray does not mean that you hurt.

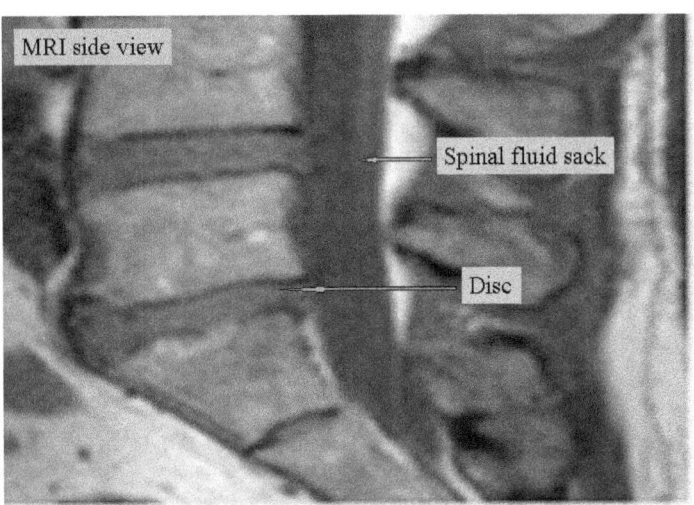

Figure 2. MRI from a side view. Note that when compared to a plain X ray that you can see more anatomic structures.

Bone scanning is done using a technetium isotope tracer injected into a vein. This tracer is distributed according to the bone blood flow. A greater blood flow to the bone from trauma such as a fracture or arthritis is compatible with greater bone absorption of the tracer. Total body radiation occurs but is low following a bone scan. The three-phase bone scan consists of the administration of a radioactive tracer followed by scanned images on three occasions. The first image is phase 1 Phase one measures blood flow on the first pass of the tracer. The second phase assesses the blood vessel system while the third phase assesses the turnover of bone, which can be seen in fractures or tumors. Bone scans are frequently used to diagnose RSD.

Electromyography (EMG) and the Nerve Conduction Velocity Tests (NCV) (figure 3) are two diagnostic tools that are helpful to the pain management doctor. These two tests allow the assessment of the location, the pathogenesis, and the prognosis of neuromuscular lesions. Loss of the outside wrapper (myelin) of a nerve or nerves is assessed by the nerve conduction velocity test. Abnormalities of a nerve take 3-5 days to develop. An EMG is a needle test to determine if your muscle is diseased or injured.

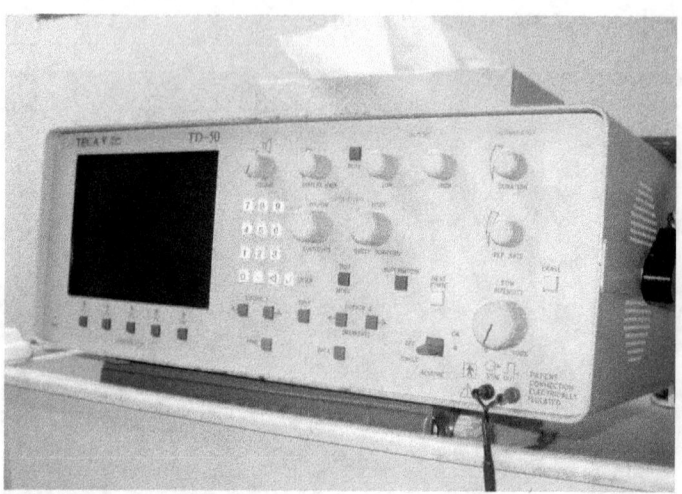

Figure 3. EMG/NCV machine.

Abnormalities in your muscle can take five to six weeks to become evident. Muscles that are closer to your brain manifest electrophysiological abnormalities sooner than more distal muscles. Focal defects in your nerve may cause NCV slowing across the defect. A NCV measures how fast your nerve sends and impulse. Stimulation of the nerve is done at one end of your nerve and the velocity is measured at another end of your nerve. Generalized nerve pathology results in a reduced nerve conduction velocity. In other words your nerve impulses are slower than normal.

Electromyography (EMG) measures the response of muscles and nerves to electrical activity. It's used to help determine muscle conditions that might be causing muscle weakness, including muscular dystrophy and nerve disorders. A needle electrode is inserted into your muscle (the insertion might feel similar to a pinch) and the signal from the muscle is transmitted from the electrode through a wire to a receiver/amplifier, which is connected to a device that displays readout. EMGs can be uncomfortable and scary to kids, but aren't usually painful. Occasionally kids are sedated while they're done.

Distal latency is the assessment of the distal conduction velocity of your painful nerve that can be affected by the neuromuscular junction that is the location where the nerve and muscle join. Some muscle diseases may have normal NCV studies but electromyographic (EMG) abnormalities usually occur in these situations. EMG measures muscle electrical activity. A reduction in the size of the waves on an oscilloscope (a screen with waves that move across the screen) is proportional to the nerve loss to the muscle. It should be noted when a muscle is penetrated by an EMG needle, the normal muscle is quiet when it is at rest.

Muscle fiber firing at the time of needle insertion can give your doctor an indication of any muscle disease. The NCV assesses the speed at which your peripheral nerves transmit electrical signals. Your nerve is stimulated, usually with surface electrodes, which are patch-like electrodes placed on your skin over the nerve at various locations. One electrode stimulates the nerve with a very mild electrical impulse. The other electrode records the resulting electrical activity. The distance between electrodes and the time it takes for electrical impulses to travel between electrodes are used to calculate the nerve conduction velocity.

A muscle should contract when stimulated by a nerve impulse. A needle electrode is inserted through the skin into your muscle. There should be a short burst of electrical activity at this time. The electrical activity detected by this electrode is displayed on an oscilloscope, and may be heard through a speaker. After placement of the electrodes, you may be asked to contract certain muscles. The presence, size, and shape of the waveform make up an action potential. This waveform provides information about the ability of your muscle to respond to electrical stimulation. These tests are useful for investigating nerve and muscle function in diseases such as peripheral neuropathy, compression neuropathy etc.

7. Qualifications to Practice Pain Management

Did you know that the doctor doing your injection may have trained at a weekend course or may have learned on the job? The weekend courses are cadaver courses. Do you want to be that doctor's first patient? You might be. There are no regulations on pain injectionists.

Doctors who manage pain are frequently anesthesiologists. Anesthesiologists ensure that you are safe, pain-free and comfortable during and immediately following surgery. But not everyone realizes that decades of research and work done by anesthesiologists have led to the development of newer, more effective treatments for patients who have pain unrelated to surgery. Many techniques used to make surgery and childbirth virtually painless is now being used to relieve other types of pain. In fact, the work pioneered by anesthesiologists has led to treatments for pain control outside the operating room.

Frequently an anesthesiologist heads a team of other specialists and doctors who work together to help you manage your pain. Pain medicine doctors are experts at diagnosing why you are having pain as well as treating the pain itself. Some of the more common pain problems they manage include: arthritis, back and neck pain, cancer pain, nerve pain, migraine headaches, shingles, phantom limb pain for amputees and pain caused by AIDS. Pain medicine doctors are experts at diagnosing why you are having pain as well as treating your pain.

Like other physicians, anesthesiologists have completed four years of medical school. They spent four more years learning anesthesiology and pain medicine during residency training. Many anesthesiologists who specialize in pain medicine receive an additional year of fellowship training to become an expert in treating pain. Some also have done research, and many have special certification in pain medicine through the American Board of Anesthesiology. This board is the only organization recognized by the American Board of Medical Specialties to offer special credentials in pain medicine. Medical specialty certification in the United States is a voluntary process. While medical licensure sets the minimum competency requirements to diagnose and treat patients, it is not specialty specific. Board certification demonstrates a physician's expertise in a particular specialty and/or subspecialty of medical practice. Pain medicine is a subspecialty of anesthesiology.

The American Board of Medical Specialties member boards (24) are responsible for setting the standards for quality practice in a particular medical specialty. Each Member Board has a board of trustees or directors, all of whom are certified in that Board's medical specialty. Individual Member Boards evaluate physician candidates to ascertain if the candidate completed the appropriate residency requirements and if he or she has an institutional or valid license to practice medicine. If a physician meets these basic admission standards, the Member Board will evaluate the candidate using written and oral examinations. Because specialties differ so widely, the criteria that inform these tests are quite different. What makes someone a good anesthesiologist does not necessarily make him or her a competent cardiologist.

Ultimately, the measure of physician specialists is not merely that they have been certified, but that they keep current in their specialty. The American Board of Medical Specialties requires maintenance of certification that is a formal means of measuring a physician's continued competency in his or her certified specialty and/or subspecialty. To become recertified a physician must: hold a valid, unrestricted medical license, meet educational and self-assessment programs determined by the particular Board, demonstrate specialty-specific skills and knowledge, demonstrate the use of best evidence and practices compared to peers and national benchmarks.

Unfortunately, a physician needs no credentials to practice pain management other than a medical or osteopathic medicine degree and a state medical license. As a result, there are no guidelines as to who can call themselves a pain medicine "specialist". There are no local, state or national standards with respect to pain management. The American Academy of Pain Medicine, the American Academy of Pain Management, and the American Board of Anesthesiologists administer written examinations to certify pain management doctors. These organizations provide continuing education courses annually. Some physicians do not certify through one of these organizations but classify themselves as pain physicians. They may go to a weekend cadaver course to be able to do a certain procedure. You should ascertain that you are not the first live patient that one of these doctors practices on after finishing a weekend cadaver course. Other professionals (plumbers, nurses, teachers, policemen, etc.) must have formal training to practice their profession. Pain doctors in many instances do not! Does this frighten you? It should!! I recommend that you investigate your prospective pain physician before you receive treatments that will not benefit you or may actually harm you.

Ask to see your physician's credentials. Hospitals in many instances may want anyone who will show up and has a medical degree to do potentially mutilating procedures on patients who have insurance. You should ascertain if your pain doctor has completed a fellowship (specialized training in pain medicine) or has sufficient experience through residency training to do a procedure on you. With these facts in mind, you must do your homework when choosing a pain medicine physician. Most university pain centers require that pain medicine physicians have a formal fellowship before they can begin to treat patients.

It is your duty to find the best-trained physician. Your insurance plan will list doctors approved by their plan. Some companies have strict criteria before admitting physicians to their plan. The American Association of health plans lists a Web site with a doctor finder at www.aahp.org. Click on to this site and follow the instructions to help locate a physician. A background information check can be done on every physician. Go to ama-assn.org/aps/amahg.htm. Has your physician ever been disciplined by a state medical board? Find out by going to the Web site ama-assn.org/ama/pubcategory/2645.html to find out if your doctor is a health hazard. If your doctor has been disciplined find out the reason.

A physician should have some certification from a medical specialty like anesthesiology, physical medicine etc. to do pain medicine in addition to completion of a fellowship. Ideally, the pain medicine physician should have further training in pain medicine. If your physician has no certification in any specialty you should eliminate that physician from your list of potential treating physicians. There may be some qualified doctors who have not taken a certification test but it would be extremely difficult identifying those individuals who are truly competent. You should ask your physician if he or she has credentials in pain medicine including research publications etc. To ascertain if your physician is certified by the American Board of Medical Specialties, go to the website www.ABMS.org.

8. Narcotic Drugs

Narcotic drugs are prescribed for postoperative pain, cancer pain and for some chronic pain syndromes. Narcotic drugs can relieve moderate to severe pain. The term narcotic refers to agents that benumb or deaden nerves, causing loss of feeling or paralysis. Psychodelic drugs like LSD, contrary to popular belief are not narcotics. Many law enforcement officials in the United States inaccurately use the word "narcotic" to refer to any illegal drug or any unlawfully possessed drug.

Most medical professionals prefer the term opioid which refers to natural, semi-synthetic and synthetic substances that behave pharmacologically like morphine. The Opioids are a class of controlled pain-management drugs that contain natural or synthetic chemicals based on morphine, the active component of opium. These narcotics effectively mimic the pain-relieving chemicals that the body produces naturally. Opioids are the most often prescribed pain-relievers because they are so effective.

Morphine is the standard to which other opioid drugs are compared. Morphine is frequently prescribed to alleviate severe pain after surgery. Codeine can be helpful in soothing somewhat milder pain, as are oxycodone (OxyContin, an oral, controlled-release form of the drug), propoxyphene (Darvon), hydrocodone (Vicodin), hydromorphone (Dilaudid) and meperidine (Demerol), which is used less often because of its side effects. Diphenoxylate or Lomotil can also relieve severe diarrhea, and codeine can ease severe coughs.

The primary medical use of opioids is to relieve pain. Other medical uses include control of coughs and diarrhea, and the treatment of addiction to other opioids. Opioids can produce euphoria, making them prone to abuse. Opioids should only be used for moderate to severe pain that has not responded to non-narcotic drugs like aspirin or ibuprofen.

Narcotics can be used alone like oxycodone or used in combination with aspirin, ibuprofen or acetaminophen (Tylenol). Some narcotics like oxycodone or morphine are available as an extended release tablet that must be swallowed whole. Tablets, which are not extended release, may be split..

In 1914, the Federal Government passed a law that prohibited prescribing opioid drugs for recreational use. The Federal Controlled Substances Act of 1970 formulated schedules for drugs. You need to be aware of three of five schedules; I, has no current accepted medical use like heroin or marijuana, II; high abuse and dependence potential like morphine, codeine or oxycodone, and III; includes drugs with a lesser dependence and abuse liability. Hydrocodone (Vicodin) is a schedule III drug. Valium, a relaxant is a schedule IV drug and some cough medicines are schedule V drugs. Oxycodone (Oxycotin) is a schedule II drug which means that it is potentially more habit forming than hydrocodone.

There is a difference between the descriptions of narcotic drugs and opioids. Opioids are drugs like morphine, hydrocodone etc. Narcotics are extremely addictive drugs and include heroin and other drugs that can cause sedation. Opioids act by attaching to a group of proteins called opioid receptors, found in the brain, spinal cord and gastrointestinal tract. When these drugs link to certain opioid receptors in the brain and spinal cord they can block the transmission of pain messages to the brain.

For the purposes of discussion in outlining the pharmacologic activity of these compounds, the opioids will be classified as (1) agonists, (2) antagonists, and (3) mixed agonist-antagonists. All drugs bind to receptors that exist on the outer membrane of your cells. Narcotics bind to narcotic receptors on cells in the brain and spinal cord. Opioid receptors may also be recruited on tissue cells outside of your central nervous system such as your knee following an injury. An injection of morphine into your knee may alleviate your pain.

When opioids turn on a receptor, that receptor decreases pain signals usually in your spinal cord that prevents pain signals from going to your brain. As a result, your pain perception is decreased. Experimental studies involving binding of opioids to specific receptors in the brain and spinal cord have substantiated the hypothesis that these receptors exist which mediates the actions of the opioid drugs to stop pain signals to your brain. There are two basic classes of opioid receptors called mu and kappa receptors.

Other classes exist (e.g. delta) but are not important for the discussion of your pain in this chapter. These receptors also appear to be the site of action of the endogenous (pain drugs produced by your body) opioid-like substances and have been divided into three major categories, designated mu, and kappa.

It has also been proposed that at least two subtypes of each category of opioid receptors exist. Experimental evidence suggests that activation of mu receptors (found principally at sites in the brain) is associated with analgesia, respiratory depression, euphoria, and physical dependence. The kappa receptors (located within the spinal cord) are believed to mediate spinal analgesia, constriction of the pupil size and sedation. The other receptors may influence affective behavior, and although some physicians believe that activation of these receptors plays a role in opioid-induced analgesia, this remains controversial.

Since a number of different compounds, (e.g., certain antihistamines, some steroids, and anti psychotics have phencyclidine) none of which are opioid in structure but can affect binding affinity for these sites. Agonistic (stimulating) opioids act as analgesics by binding to and activating both mu and kappa receptors in the brain and spinal cord. The opioid antagonists bind to all categories of opioid receptor sites throughout the body, but fail to activate them. These compounds are not used for pain control; rather, the utility of these drugs lies in their ability to reverse an overdose of opioids including narcotics.

The compounds that comprise the mixed agonist-antagonist group are more recent additions to the clinically important opioids. These drugs are semi-synthetic derivatives of morphine, the chemical structures of which have agonistic activity at some kappa receptors but antagonistic activity at mu receptors, e.g., pentazocine, butorphanol, and nalbuphine, or partial agonistic activity at mu receptors and antagonistic activity at kappa receptors, eg. buprenorphine. All are effective analgesics since they stimulate either mu or kappa receptors.

Chemically, the opioid agonists include a number of classes of drugs, all of which have pharmacologic effects similar to those of morphine. Morphine is the oldest known drug of this class. It remains as the prototype for the opioid group and is the standard to which all other opioid analgesic drugs are compared. Opioid drugs decrease pain but also affect all organ systems. Your pituitary gland in your brain can be adversely affected by chronic narcotic use. For example in males opioids can decrease testosterone that can cause depression and erectile dysfunction. Drowsiness and blurred vision can occur. Changes in mood can occur. An inability to concentrate can occur.

Euphoria can be experienced in 20% of individuals taking opioid drugs. Euphoria can be the cause of addiction. Opioids can stop your respiratory drive that can cause you to stop breathing. Narcotics affect your stomach by slowing down the passage of food in combination with your brain to cause nausea and vomiting. Opioids can cause a significant decrease in your blood pressure that may cause you to fall. Opioids decrease movement of the bowel resulting in constipation. Morphine can make gall bladder disease worse by contracting a valve where the gall bladder meets the intestine called the sphincter of Oddi. Opioid drugs can result in a release of histamine from certain cell in the body that can cause itching and a rash. As you can see opioid drugs can have side effects.

Tolerance, addiction and physical dependence can occur with opioid drugs. Tolerance occurs when it takes more of the drug to cause the same decrease in your pain. This is not addiction. Patients may find that they develop tolerance to opioid pain medications and may need to have their doses increased in order to be effective. Tolerance has not been shown to lead to drug addiction. Physical dependence is a condition that occurs when continued use of the drug is needed to prevent a withdrawal reaction. Steady use of opioids can result in tolerance to the drugs so that higher doses must be taken to achieve the same effects. Long-term use also can lead to physical dependence—the body adapts to the presence of the drug and withdrawal symptoms occur if use is reduced abruptly.

Addiction is an intense craving for an opioid and is often associated with recreational use. Signs and symptoms of addiction include yawning, sweating, restlessness, irritability, anxiety, nasal discharge, tearing, dilated pupils, gooseflesh, tremors, loss of appetite, body aches, nausea and vomiting, fever and chills and an increase in heart rate and blood pressure. These symptoms last 7-10 days. Minor symptoms can begin in 8-12 hours after the last dose of the opioid. The more severe symptoms like nausea and vomiting begin 48-72 hours after the last dose of the drug. With respect to agonist drugs, morphine is the prototype. It can be administered by mouth, rectum or by injection into muscle or vein. Is is prepared in a capsule, tablet or a liquid. It is available by a rectal suppository as well. This route of administration is used for those patients who cannot swallow or are having severe vomiting. Hydromorphone and oxymorphone also come in the form of rectal suppositories. The duration of action of opioids varies from drug to drug. Sustained release morphine and oxycodone give a longer duration of action. Immediate release drugs (eg. OXIR) give a faster onset but have a shorter duration of action. Fentanyl, which is 75 times more potent than morphine is available in a patch and sucker, forms. The fentanyl patch is used for severe constant pain. The pain relief is continuous. The sucker, which only comes in a raspberry flavor, is used for severe cancer pain in instances where the severe pain fluctuates. Fentora is another oral form of fentanyl.

With respect to the fentanyl pain patch, the amount of drug released is controlled by small holes in a membrane in the patch. A larger hole permits the release of fentanyl into your body. The patches are available in different doses. The fentanyl is released for 48-72 hours. Patients with a fever can be at a risk for an overdose as the amount of fentanyl administered to your body can increase by 25% for every 3^0C increase in body temperature. The advantage of the patch is that patients do not have to take frequent pills during the night. The patch should be applied to a hairless surface.

Darvon (propoxyphene) and codeine are weaker opioids that are used to treat mild pain. They may be combined with acetaminophen to make each more potent. You need to be aware that smoking tobacco can decrease the potency of Darvon and hydrocodone.

Tramadol (Ultram) is an interesting drug and may be used for moderate to moderately severe pain. It has a low abuse potential. It is not a scheduled drug. It activates mu and kappa receptors. The side effects are minimal when compared to opioid drugs. Tramadol does not produce withdrawal symptoms like opioids. The advantage of tramadol over other drugs is that tramadol inhibits noreoinepherine and serotonin. These two substances in the brain and spinal cord also decrease pain. The opioid drugs do not have this effect. Tramadol can cause nausea dizziness and headaches. Tramadol does not lower the heart rate or blood pressure. Tramadol provides pain relief similar to codeine and propoxyphene. Naloxone and naltrexone are drugs that reverse the respiratory effects of opioids. Naltrexone can be given orally. The only time that these drugs are given is to treat opioid intoxication. Butorphanol (Stadol) and pentazocine (Talwin) are called mixed agonist-antagonists drugs. These drugs show receptor selectivity and these two drugs stimulate kappa receptors. These drugs have less opioid abuse tendencies than the agonist drugs. Opioids on the other hand work on both mu and kappa receptors. Strong opioids exist which are usually reserved for cancer patients or other patients with severe pain.

Hydromorphone (Dilaudid) and levorphanol (Levo-Droman) are eight and five times more potent than morphine. Meperidine (Demerol) is an opioid that is weaker than morphine. It is used infrequently in pain management as it can cause tremors or seizures if used on a chronic basis. Methadone is a synthetic drug similar to morphine. The advantage of methadone for your pain management is that it does not cause euphoria. Methadone however, can cause a conduction problem in your heart. Consequently, patients have died from heart problems after being prescribed methadone. Hydrocodone and oxycodone are two opioids used for moderate to moderately severe pain. These drugs are usually combined with aspirin and acetaminophen which can potentiate the analgesic efficacy of these drugs.

Another fact that you need to know is that opioid drugs can actually cause you to experience increased pain. This observation is called opioid induced pain. Many physicians are unaware of this fact. In this situation, a reduction in your dose of your medicine or stopping it can actually decrease your pain. This phenomenon can also be seen in patents who have spinal morphine drug delivery systems.

As one can see, there are many opioids that can be used for the management of your acute chronic pain. The proper choice of your medication is dependent upon the magnitude of your pathology, the side effects of the drug prescribed, the effectiveness of the drug and your overall health.